Ceramic Armor and Armor Systems

Related titles published by The American Ceramic Society

Ceramic Armor Materials by Design (Ceramic Transactions, Volume 134)
Edited by J .W. McCauley, A. Crowson, W.A. Gooch Jr., A.M. Rajendran, S.J. Bless, K.V. Logan,
M. Normandia, and S. Wax
©2002, ISBN 1-57498-148-X

Other titles of interest

The Magic of Ceramics
David W. Richerson
© 2000, ISBN 1-57498-050-5

Advances in Ceramic Matrix Composites VIII (Ceramic Transactions Volume 139)
Edited by J.P. Singh, Narottam P. Bansal, and M. Singh
©2002, ISBN 1-57498-154-4

Innovative Processing and Synthesis of Ceramics, Glasses, and Composites VI (Ceramic Transactions Volume 135)
Edited by Narottam P. Bansal and J.P. Singh
©2002, ISBN 1-57498-150-1

Innovative Processing and Synthesis of Ceramics, Glasses, and Composites V (Ceramic Transactions Volume 129)
Edited by Narottam P. Bansal and J.P. Singh
©2002, ISBN 1-57498-137-4

Advances in Ceramic Matrix Composites VII (Ceramic Transactions Volume 128)
Edited by Narottam P. Bansal, J.P. Singh, and H.-T. Lin
©2001, ISBN 1-57498-136-6

Advances in Ceramic Matrix Composites VI (Ceramic Transactions Volume 124)
Edited by J.P. Singh, Narottam P. Bansal, and Ersan Ustundag
©2001, ISBN 1-57498-123-4

Innovative Processing and Synthesis of Ceramics, Glasses, and Composites IV (Ceramic Transactions Volume 115)
Edited by Narottam P. Bansal and J.P. Singh
©2000, ISBN 1-57498-111-0

Innovative Processing and Synthesis of Ceramics, Glasses, and Composites III (Ceramic Transactions Volume 108)
Edited by J.P. Singh, Narottam P. Bansal, and Koichi Niihara
©2000, ISBN 1-57498-095-5

For information on ordering titles published by The American Ceramic Society, or to request
a publications catalog, please contact our Customer Service Department at:

Customer Service Department
735 Ceramic Place
Westerville, OH 43081, USA
614-794-5890 (phone)
614-794-5892 (fax)
info@ceramics.org

Visit our on-line book catalog at www.ceramics.org.

Ceramic Transactions
Volume 151

Ceramic Armor and Armor Systems

Proceedings of the Ceramic Armor and Armor Systems symposium held at the 105th Annual Meeting of The American Ceramic Society, April 27-30, 2003, in Nashville, Tennessee

Edited by

Eugene Medvedovsk
Ceramic Protection Corp.

Published by
The American Ceramic Society
735 Ceramic Place
Westerville, Ohio 43081
www.ceramics.org

Proceedings of the Ceramic Armor and Armor Systems symposium held at the 105th Annual Meeting of the American Ceramic Society, April 27-30, 2003, in Nashville, Tennessee.

COVER PHOTO: "SEM of a TiO$_2$ coating structure" is courtesy of R. Gadow and K. von Niessen and appears as figure 6 in their paper "Lightweight Ballistic Structures Made of Ceramic and Cermet/Aramide Composites" which begins on page 3.

For information on ordering titles published by The American Ceramic Society, or to request a publications catalog, please call 614-794-5890.

4 3 2 1–06 05 04 03
ISSN 1042-1122
ISBN 1-57498-206-0

Contents

Ballistic Testing Study and Ballistic Performance of Ceramic Armor and Armor Systems

Preface

Reliable ballistic protection of military and police personnel, equipment, vehicles, aircraft and helicopters is presently generally impractical without use of ceramic-based armor systems. The development and manufacturing of ceramic armor and armor systems have received significant attention by both ceramic manufacturers and military specialists.

The international symposium Ceramic Armor and Armor Systems was held during the 105th Annual Meeting of The American Ceramic Society, April 27-30, 2003 in Nashville, Tennessee. This symposium brought together scientists and engineers working with the development and manufacturing of armor ceramics, ceramic-based armor systems, as well as studies of fracture mechanisms and ballistic evaluations of ceramic-based armor systems. A total of 30 papers, including 12 as invited, were presented by the leading specialists from 8 countries (Canada, Germany, Israel, Korea, Russia, Ukraine, United Kingdom and the United States). The speakers represented universities, government research centers and laboratories and industry. The symposium attracted many ceramic specialists from The American Ceramic Society and also armor design and ballistic specialists from many countries.

These proceedings contain 15 invited and contributed papers presented and discussed at the symposium. The papers describe results of the latest achievements in the area of ceramic armor systems devoted to ceramic armor design and modeling, ceramic armor materials and composites development and manufacturing, physical properties and structures of armor ceramics, fracture mechanisms of armor ceramics and composites, and ballistic testing and performance of ceramic armor systems. The papers also consider new tasks and approaches in the area of armor ceramics and armor systems. Each manuscript presented was reviewed in accordance with The American Ceramic Society's review process.

As the organizer of the symposium and the editor of this proceedings, I am grateful to the session chairs (Drs. Rainer Gadow, Dale Niesz and Victor Greenhut) and to all the participants for their contribution and co-operation, time and effort, to all reviewers for their comments and suggestions and to the ACerS staff of the meetings and publication departments for their assistance. The financial support of The American Ceramic Society is gratefully acknowledged.

I hope that this volume will be good addition to the recently published literature related to ceramic armor, such as the proceedings of the armor symposiums held at the PAC RIM Conference (Hawaii, 2001) and the 27th Cocoa Beach Conference, 2003, also conducted by The American Ceramic Society. This volume should be of interest to researchers and engineers working with all aspects of ceramic armor systems. The results described herein will help in the development and implementation of advanced ceramic armor with improved performance and finally in protecting human life.

Eugene Medvedovski

Ceramic Armor Materials Development

LIGHTWEIGHT BALLISTIC STRUCTURES MADE OF CERAMIC AND CERMET / ARAMIDE COMPOSITES

R. Gadow and K. von Niessen
Institute for Manufacturing Technologies of Ceramic Components
and Composites (IMTCCC/IFKB), University of Stuttgart
Allmandring 7b, D-70569 Stuttgart, GERMANY

ABSTRACT

Ceramic and cermet coatings on fiber fabrics should enhance their performance in ballistic protection. Based on thermal spray technologies a coating process for hard material layers even on temperature sensitive fiber substrates has been developed, so that the coated fabrics retain their flexibility. High speed and high rate cermet and ceramic coating is performed with simultaneous cooling in order to apply thick, hard and refractory cermet and oxide ceramic coatings can be applied on lightweight aramide fabrics without damaging the initial fibers. The hard material, aramide composite fabric combines the advantages of aramide fabrics and hard, refractory materials. A fully flexible, highly tenacious and lightweight fabric with a hard and refractory top coating is developed. The penetration of bullets, knives and blades through such hard material coated multilayer fabrics is effectively prevented.

INTRODUCTION

Ballistic protection is required for personal use, vehicles and permanent structures which are subject to ballistic threats.[1,2] Military and civilian ballistic protection is divided into flexible lightweight protection and massive, stiff armor. Lightweight ballistic protection can be made of flexible aramide fibers and is primarily used as body armor[3,4] (see Fig. 1).[5] Stiff armor consists of multilayer steel as well as dense bulk ceramic plates and stiff fiber reinforced materials (see Fig. 2).[6] The main disadvantages of solid and stiff armor are its heavy weight and inflexibility. For personnel protection as well as protection of aircrafts and cars only light and flexible materials can be used.[7] Light and flexible fabrics made of aramide or other high tenacity fibers meet some of these demands, but their protection is not sufficient. Sharp blades and high speed bullets can pierce these fabrics even if several layers are used. The state of the art solution to protect body armor against knives and bayonets is the extra use of layers made of titanium foil (see fig. 3) or resin treated fabrics. But this significantly increases the specific weight of the body armor.[8] On the market there is only one stab protection solution that follows the idea of hard materials on top of a aramide fabric. Produced by Twaron© it is called Twaron© SRM. Silicon carbide particles are dispersed in a polymer matrix and glued on top of an aramide fabric.

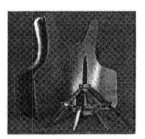

Fig.1 Personnel protection made of fiber fabrics[8]

Fig.2 Heavy armor made of bulk ceramic[5]

Fig.3 Extra protection made of titanium foil[8]

Lightweight engineering in advanced product development is mainly based on composite technologies. This paper focuses on a new approach by coating fabrics made of high tenacity aramide fibers with refractory cermet and oxide ceramics of high hardness by thermal spray technologies without damaging the initial fibers. The developed composite made of aramide fabrics, cermet and ceramics retains its flexibility. The well established aramide fiber fabric multilayer structure for personnel textile protection systems can be distinctly improved by these additional coatings.

The combination of high tenacity fiber woven fabrics and high performance hard material coatings prevent penetration by most bullets, knives and blades effectively. The high tenacity aramide fabric ensures the impact resistivity by damping and dissipating the shock wave velocity. The ceramic coating increases the fiber-to-fiber friction which prevents wave distortion and delamination. Penetrating objects cannot change the fabric structure and push the fibers aside. The hard material coatings blunt sharp metal blades by abrasion so they cannot cut the fabric, and the high friction between the ceramic coating and the metal blade stops further penetration.

MATERIAL SCREENING

The material screening focusses on the use of high tenacity aramide fiber fabrics and hard and highly refractory cermet and oxide ceramics. As aramide fabric the commercially available fabric Twaron© CT 710 (Teijin Twaron GmbH, Wuppertal, Germany), which is standard for ballistic protection, has been selected. The material properties of its fibers is summarized in table I.

Table I. Properties of Twaron© fibers[9]

Fiber fabric	Density ρ [g/cm^3]	Tenacity σ [MPa]	Initial modulus E [GPa]	Decomp. temp. T_D [°C]	Specific heat C_P [J/kgK]	Max. appl. Tem. T_M [°C]
Twaron© CT 710	1.45	2,800	85	500	1420	200

Due to their high hardness and wear resistance the cermet WC Co 83/17 as well as the oxide ceramics Al$_2$O$_3$ and TiO$_2$ have been chosen as coating materials for thermal spraying. To improve the bonding strength of the ceramic coatings on the fabric, AlSi 12 is used as additional bond coat. The bulk material properties of the coating materials are shown in table II.

Table II. Bulk material properties of coating materials[10, 11, 12]

Oxide ceramic	Density ρ [g/cm^3]	Vickers hardness H_V [-]	Youngs modulus E [GPa]	Melting temp. T_M [°C]	Specific heat C_P [J/kgK]
Al$_2$O$_3$	3.98	2,200	400	2,047	1,047
TiO$_2$	4.25	1,150	205	1,860	730
AlSi 12	2.75	205	102	660	880
WC	15.7	2,200	669-696	2,800*	-
Co	8.8	250	211	1,493	440
WC Co 83/17	14,1	1,850	560	2,800 / 1,410	-

* decomposition temperature

In order to apply these hard materials by thermal spraying they have to be available as spray powders. After a sintering process the used powders are mechanically broken and milled to a grain size of 10 – 22 μm.

DEPOSITION OF OXIDE CERAMIC COATINGS ON LIGHTWEIGHT FIBER FABRICS BY THERMAL SPRAYING

The thermal spray process allows the application of a broad variety of metallurgical, cermetic and ceramic coatings on a variety of substrates. A key feature of the thermal spraying technique compared to many other methods is the substrate's low thermal load during the coating process. By using simultaneous air or liquid CO_2 cooling techniques, the substrate's temperature can be kept relatively low e.g. between 50° and 150°C. The Atmospheric Plasma Spray (APS) process uses an electric arc discharge between a water cooled copper anode and a tungsten cathode as an energy source. This electric arc discharge dissociates and ionizes the working gas and builds up a plasma expanding into the atmosphere forming a plasma gas jet (see Fig. 4).[13]

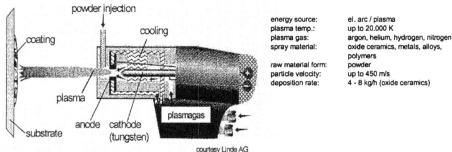

Fig. 4 The Atmospheric Plasma Spray (APS) process[13]

The spray powder, suspended in a carrier gas, is injected into the heat source of the torch. After being totally or partially molten and being accelerated, the powder particles impact on the substrate's surface where they are quenched and solidified within 10^{-5} to 10^{-7} second[14]. The coating's build-up is the result of the molten powder particles impacting one upon the other (see figs. 5 and 6).

Ceramic Armor and Armor Systems

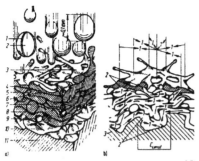

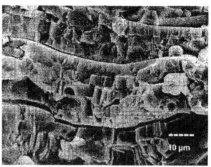

Fig. 5 Lamellar coating build up [15]

Fig. 6 SEM of a TiO_2- coating structure

During the atmospheric plasma spraying process temperatures up to 20,000 °C are obtained. This process is therefore mainly used for deposition of refractory materials like oxide ceramics. For the application of thermally sprayed coatings on fiber woven fabrics the torch movement is performed by a 6-axis robot system and a metal frame is used to insert and tighten the samples. A steel wire cloth within the frame supports the flexible fabric structure during the coating process and allows cooling because of its open and permeable design. The meandering movement and the metal frame are shown in fig. 7.

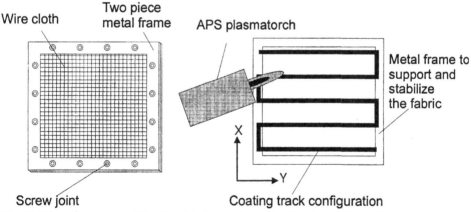

Fig. 7 Mounting support for the fabrics and coating track configuration

In order to limit the thermal load on the fabrics a simultaneous cooling with compressed air or liquid CO_2 is used. Cooling nozzles are attached on both sides of the spraying torch. In addition, the process is supervised by an infrared camera

(Varioscan InfraTec ID, Dresden, Germany) and in that way the temperature of the coated samples can be controlled in real time. Fig. 8 shows a typical IR- picture during the coating process.

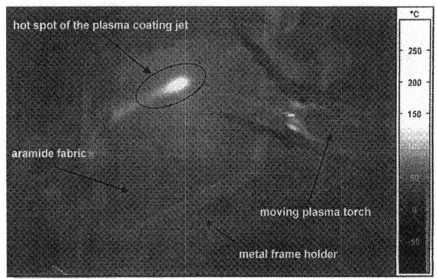

Fig. 8 IR- picture of the temp. distribution during the coating process

Because of the high relative speed between the plasma coating jet and the fabric and the intensive cooling, the thermal load on the fabric is drastically reduced. A hot spot of only 190 °C is measured.

MECHANICAL CHARACTERIZATION

With regard to the use of the coatings for lightweight ballistic protection the main focus of the characterization is on the determination of weight, puncture resistance, hardness and wear resistance as well as on the evaluation of the coating's bonding strength on the first fiber layers. During the coating build-up of thermally sprayed layers, porosity and microcracks cannot be avoided. For the coating of flexible fabrics the formation of porosity and microcracks in the coating is desired because it leads to a higher flexibility of the fabric. But if the porosity is too high, the hardness and other mechanical properties of thermally sprayed coatings decrease. So a balance between porosity and mechanical properties has to be found.

The thickness of the hard material coatings on the fabric is in the range 75 +/- 5 µm. Fig 9 shows a schematic drawing of the intended structure of the coated fabric.

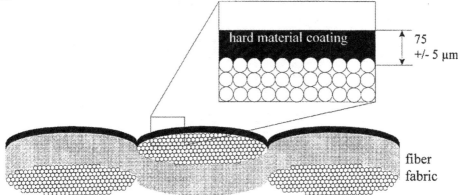

Fig. 9 Intended structure of the hard material coated fabric[16]

Fig. 10 shows a cross section of a Twaron© fabric coated with an Al_2O_3 oxide ceramic layer. The lamellar structure and the good wetting behavior of the ceramic coating on the first layers of the fabric are visible. The macro-and the micro-structure of the coated fabric`s surface is typical of thermally sprayed coatings (see fig. 11). The structure of the fabric is still visible in the macro-structure. Even though the TiO_2- and Al_2O_3- coatings have melting points above 1800° and 2000°C respectively, there is no significant polymer fiber damage.

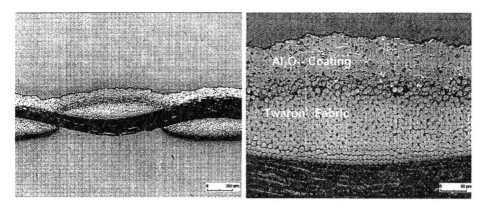

Fig. 10 Cross section of a thermally sprayed Al_2O_3 coating on a Twaron© fabric

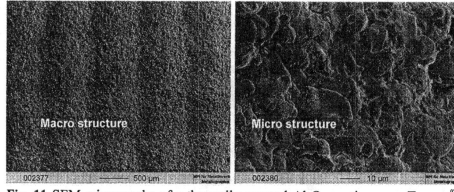

Fig. 11 SEM micrographs of a thermally sprayed Al_2O_3 coating on a Twaron[©] fabric

In order to evaluate the coating quality metallographic examinations have been performed. The coating porosity determined by an image analysis is expressed by the relative pore volume content V_P [%]. An automized universal hardness indenter equipment (Fischerscope TM HCU) with a load of 500 mN is used to determine the coating hardness $H_V0,05$. For measuring the hardness of an individual fiber, the load was reduced to 10 mN (H_V 0,001). Table III and table IV show the measured porosity and hardness characteristics of the thermally sprayed coatings and of the fibers, respectively.

Table III. Measured coating porosity and microhardness (H_V 0,05)

Coating	V_P [%]	H_V 0,05
Al_2O_3	5.8	1,240 +/- 300
TiO_2	3.2	1,100 +/- 110
Al_2O_3/TiO_2	4.1	1,025 +/- 180
WC Co 83/17	2.7	1.266 +/- 188
AlSi 12	1.44	138 +/- 10

Table IV Microhardness of individual fibers (H_V 0,001)

Fiber	H_V 0,001
Twaron[©] CT 710	51.52 +/- 7

One of the intentions of ceramic coatings on fiber fabrics is to blunt metal blades or other penetrating objects by abrasion. In order to judge the wear behavior of a metal counterpart on the cermet and oxide ceramic coatings, dry-running-oscillating-pin-on-disc tests are performed. As a counterpart 100Cr6 balls with a hardness of 1.165 $H_V0.05$ and a diameter of 5 mm are used. The selected number of oscillating strokes is 10.000, the sliding velocity is 70 mm/s, the length of strokes is 5 mm with an imposed normal load of 10 N. It can be assumed that a higher volumetric loss on the 100Cr6 ball after the end of the tribological tests indicates a better ability of the coating to blunt a metal blade. Non coated fabrics have also been tested in order to estimate the abilities of the fabric itself to blunt a metal blade,. Fig. 12 shows the volume loss of the 100Cr6 balls with different coatings.

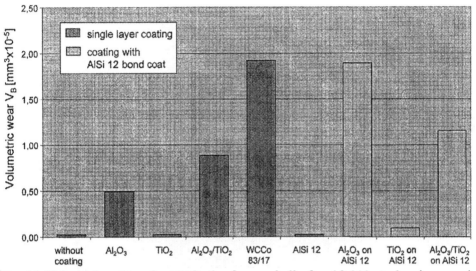

Fig. 12 Volume loss V_B of a 100Cr 6 reference ball after 10.000 strokes in reversing slide contact with ceramic coated Twaron© Fabrics

The rather smooth fabric Twaron© does not cause any severe wear without coating. The Al_2O_3 and Al_2O_3/TiO_2 single layer coatings on Twaron cause a significant volume loss, which is much further increased by using an additional AlSi 12 bond coat. Especially the Al_2O_3/AlSi 12 multilayer coating and the WC Co 83/12 single layer coating perform well. But since the microhardness of the AlSi 12 bond coat is low and thus it does not have a noticeable impact on the abrasion behavior, only the increased higher bonding strength of a multilayer coating explains this result. Because of the protective function of the bond coat, the hard oxide ceramic coatings cannot easily be sheared. The same reason might be responsible for the good performance of WC Co 83/12. Hard WC particles guarantee a high amount of wear to the counterpart and the ductile Co matrix a low wear to the fabric. The comparatively hard TiO_2 coating has no significant influence on the volume loss, even with an AlSi 12 bond coat. This might be due to the solid lubricant properties of TiO_2, which have been described in other investigations.[17-19]

For industrial application the coating must also show a sufficient bond strength to the fabric. The coating's adhesion to the fabric is investigated on a Zwick Z100 universal mechanical testing machine by pull testing. The coated fabric samples are glued to a metal plate and a steel tension rod is glued to the coating surface by using an adhesive. After mounting the samples into the testing machine the tensile load is continuously increased. As soon as a delamination of the coating occurs, the tension load is measured and the bonding strength is determined. As the bonding strength of the coatings is limited by the maximum shear strength of the first fiber layers which are in contact with the coating, the fabrics are also tested without any coating. In this case the tension rod is glued directly on top of the fabric. Fig. 13 shows the measured bonding strengths for the used fabrics with or without AlSi 12 bond coat.

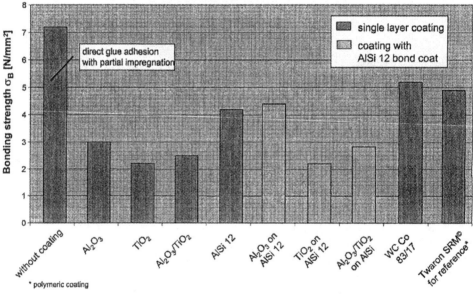

Fig. 13 Bonding strengths of the thermally sprayed coatings on fiber fabrics

The results of the experiments with non–coated fabrics show the maximum possible bonding strength a coating could reach on the fabrics due to the maximum shear strength of the fibers which are in direct contact with the coating. This maximum bonding strength of Twaron© is about 7 N/mm². None of the oxide ceramic coatings reaches this limit, but using a bond coat, the bonding strength is increased. Especially the Al₂O₃/AlSi 12 and the WC Co 83/17 coatings show a high bonding strength and the highest microhardness. The bonding strength of WC Co 83/17 is even slightly higher than the Twaron© SRM reference composite. All coatings delaminated at the fiber- coating interface.

In order to create a flexible lightweight composite, the weight gain of the fabric during coating is of great importance. The weight gain is determined by thickness, porosity and density of the coating materials. A coating thickness of 75 μm for single layer and top coatings and 30 μm for the bond coat is chosen to find an optimum between a thick coating with high mechanical properties and a thin coating with low weight and higher flexibility. Fig. 14 shows the weight gain due to the APS coating process.

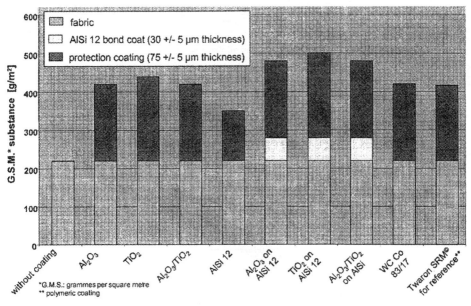

Fig. 14 Substance of the fabrics and weight gain after APS coating

The substance of the coated fabrics increases from 220 g/m² to 350 g/m² up to 500 g/m² depending on the used coating system. The multilayer coatings with a 30 μm AlSi 12 and 75 μm top coat have the highest weight gain. The cermet single layer coating WC Co 83/17 reaches a substance of only 450 g/m^2, the multilayer oxide ceramic coating Al$_2$O$_3$/AlSi 12 a substance of 480 g/m^2. But both coatings are above the reference composite Twaron© SRM with 400 g /m^2. Because of the good results obtained with Al$_2$O$_3$/AlSi 12 and WC Co 83/17 coatings, comparative stab resistance tests on these coatings are performed in comparison to uncoated Twaron© fabrics and Twaron© SRM. German standard engineered test blades K1 (A. Eickhorn GmbH, Solingen, Germany) for stab resistance tests are mounted into a Zwick Z100 universal mechanical testing machine. These blades are made of cold-rolled, hardened and tempered steel with a hardness of 52-55 Rockwell C.[20] Single layer fabrics are fixed in a specific device by hydraulic pressure to obtain a defined prestress. After that the blade stabbed the fabric with a defined velocity and the needed load is measured. The resulting puncture resistance is calculated in work [N*mm] per penetration depth [mm]. 6 stabs are carried out in one experimental run, on different single layer samples of the same fabric, using one test blade to evaluate the blunting of the blades. The typical run of the curves show an increase of the puncture resistance for every new stab using the same blade. This increase is caused by the blunting of the blade during penetration. Fig. 15 shows a measured puncture resistance curve of WC Co 83/17 coated Twaron© fabrics compared to uncoated Twaron© CT 710. The test velocity of the blade is varied from 50 to 1500 mm/min, but no influence on the results is observed.

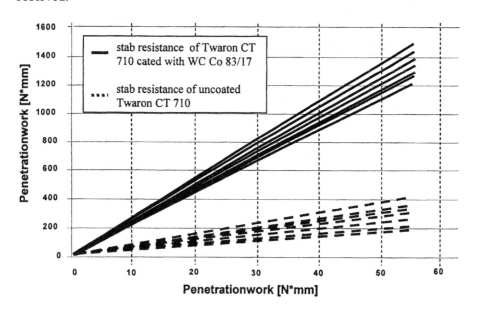

Ceramic Armor and Armor Systems

Fig. 15 Stab resistence of single layer WC Co 83/17 coated compared to uncoated Twaron© fabric

The penetration work of the coated fabrics is 5 times higher compared to the uncoated fabrics. The stab resistance of WC Co 83/17, Al₂O₃/AlSi 12 and the reference Twaron© SRM is shown in fig. 16. All hard material coatings reach a penetration work of over 1000 Nmm. The WC Co 83/17 cermet coating reaches even 1200 Nmm at the first stab and 1500 Nmm after only 6 stabs, carried out with one blade. The reference Twaron© SRM fabric shows only a stab resistance of 1040 up to 1220 Nmm.

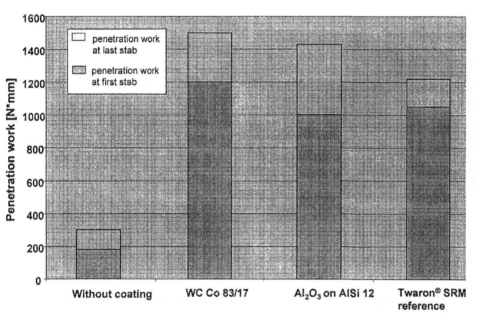

Fig. 16 Stab resistence of single layer WC Co 83/17, Al₂O₃/AlSi 12, Twaron SRM and uncoated Twaron© CT 710 fabric

The extensive blunting of the blades after only 6 stabs, which can be assumed by the increase of the stab resistance, can explicitly be seen in SEM micrographs (see fig. 17 and 18).

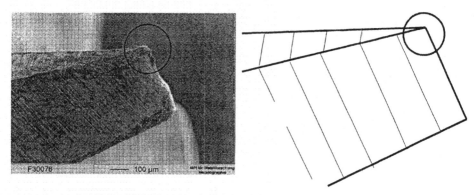

Fig. 17 SEM of a copped and sharp edged reference test blade K1 prior to testing

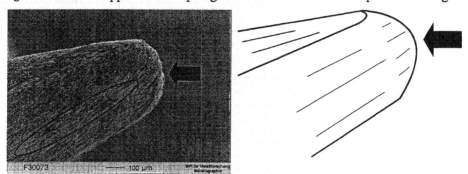

Fig 18 SEM of a test blade K1 after 6 punctures through a single layer of WC Co 83/17 coated aramide

The steel test blades with a hardness of 52-55 (Rockwell C), which started with copped and sharp edges and a plane surface turned to be out of shape, scratched and twisted after only 6 punctures. Even the scratches of individual tungsten carbide particles can be seen (fig.18).

CONCLUSIONS

The approach to combine highly refractory cermet and oxide ceramic coatings with high modulus lightweight fiber fabrics for ballistic protection has successfully been demonstrated. Atmospheric plasma spraying with well defined parameter sets and simultaneous cooling is a suitable process for the coating of cermet and oxide ceramics on top of fiber woven fabrics for ballistic protection. Even though the TiO_2- and Al_2O_3- coatings have melting points above 1800° and

Ceramic Armor and Armor Systems

2000°C, respectively, the thermal load on top of the fabrics could be limited to 190 °C maximum and so no significant polymer fiber damage has been observed. The adherent coatings remain flexible and reach a hardness up to 1260 H_V 0,05. The bonding strength is sufficient and mainly limited by the maximum shear strength of the fibers. The adhesion of the coatings and the high cycle flexibility can be improved by using metallurgical bond coats. So far the best results have been reached with a WC Co 83/17 cermet and an Al_2O_3/AlSi 12 multilayer coating. They have the highest microhardness and the highest bonding strength. Stab resistance tests were carried out on WC Co 83/12 cermet and Al_2O_3/AlSi 12 multilayer coated Twaron© fabrics and the penetration work was increased by a factor of five compared to the uncoated Twaron© fabric. A single layer of WC Co 83/17 cermet coating reaches a penetration work of 1200 Nmm at first stab, up to 1500 Nmm after only six stabs using one test blade.

Further efforts will focus on the optimization of the ratio between weight gain and increased mechanical properties like adhesive bonding strength and stab resistance. Results of first ballistic impact tests to evaluate an influence of the coatings to the ballistic performance of the fabric will be published soon.

REFERENCES

[1] C. L. Segal, "High- performance organic fibers, fabrics and composites for soft and hard armor applications," *23rd International SAMPE Conference,* USA, 651-660 (1991)

[2] N. Laetsch, "The use of Kevlar para- aramide fiber in ballistic protection garments," Du Pont de Nemours International SA, *Technical Textiles International*, Elsevier Science Ltd Oxford England, 26, (1992)

[3] J. van Dingenan and A. Verlinde, "Nonwovens and fabrics in ballistic protection," *Technical Textiles International*, Elsevier Science Ltd Oxford England, 10-13, (1996)

[4] P. G. Riewald, H. H. Yang, W. F. Shaughnessy, "Lightweight Helmet From A New Aramide Fiber," *Du Pont Fibers*, Wilmington, (1991)

[5] European Aeronautic Defense and Space Company EADS *www.army-technology.com* (Dec. 2000)

[6] J. Weatherall, M. Rappaport, J. Morton, "Outlook for advanced armor materials," *National SAMPE Technical Conference Advanced Materials: Looking Ahead of the 21st Century*, 1070-1077, USA, (1990)

[7] J-P. Charles, D. Guedra- Degeorges, "Impact Damage Tolerance of Helicopter Sandwich Structures," *Aerospatiale*, France (1999)

[8] Security Sicherheitstechnik GmbH *www.policeshop.de* (Dec. 2000)

[9] "Product data sheet Twaron", *Twaron Products*, D- 42097 Wuppertal, Kasinostraße 19-21, (1995)

[10] C. Friedrich, G. Berg, E. Broszeit, C. Berger: "Datensammlung zu Hartstoffeigenschaften," *Materialwissenschaft und Werkstofftechnik*, Vol. 28, No. 2, (1997)

[11] M. Bauccio, Ed. *ASM Engineered Materials Reference Book*, Second Edition, ASM International, Materials Park, OH, 1994.

[12] R. Kieffer, P. Schwarzkopf, F. Benesovsky, W. Leszynski, *Hartstoffe und Hartmetalle*, Springer- Verlag, Wien (1953)

[13] *Das Verfahrensspektrum beim thermischen Spritzen*, Linde AG, Werksgruppe technische Gase, Höffriegelskreuth (1990)

[14] L. Pawlowski, *The science and engineering of thermal spray coatings*, John Wiley and sons, Chichester (1995)

[15] G. Pursche, *Oberflächenschutz vor Verschleiß*, Verlag Technik, Berlin (1990)

[16] Gadow, R.; v. Niessen, K.: „Ceramic Coatings on Fiber Woven Fabrics" 26th Annual Conference on Composites, Advanced Ceramics, Materials, and Structures: A, 13 – 18.01. 2002, Cocoa Beach, USA, 2002, ACerS, *Ceramic Engineering and Science Proceeding Vol. 23, Issue 3*, Eds. H-T Lin, M. Singh, S. 277 – 285; ISSN 0196-6219 (2002)

[17] M. Buchmann, R. Gadow, D. Scherer, "Mechanical and tribological Characterization of TiO$_2$ based multilayer coatings on light metals," *The 25th Annual International Conference on Advanced Ceramics and Composites*, Cocoa Beach, 21-26, (2001)

[18] A. Skopp, M. Woydt, "Ceramic and Ceramic Composite Materials with Improved Friction and Wear Properties," *Tribology Transactions*, 48th Annual Meeting in Calgary, Alberta, Canada, Germany, 17-20, (1993)

[19] M. Woydt, "Werkstoffkonzepte für den Trockenlauf", *Tribologie und Schmierstofftechnik*, (1997)

[20] *Technische Richtlinien Schutzwesten des Unterausschusses Führungs- und Einsatzmittel* (UA FEM), Polizeitechnisches Institut (PTI) der Polizei- Führungsakademie (PFA), Germany, Anlage 6, (2000)

SILICON CARBIDE-BASED CERAMICS FOR BALLISTIC PROTECTION

Eugene Medvedovski

Ceramic Protection Corporation
3905 – 32nd Street N.E., Calgary, Alberta, T1Y 7C1, Canada

ABSTRACT

The development of lightweight and inexpensive ceramic armor is under ongoing consideration by both ceramic armor manufacturers and armor users. Various structural ceramics used for ballistic protection are reviewed as to physical properties, manufacturing and cost. Different silicon carbide-based ceramics prepared using reaction-bonding approaches or by pressureless sintering demonstrate a high level of physical properties and high ballistic performance coupled with low density. Newly developed silicon carbide-based ceramics in the SiC-Al$_2$O$_3$ system demonstrate a remarkable level of properties and high ballistic protection. The manufacturing of this type of ceramics is relatively inexpensive. The main properties of SiC-based ceramics, which affect ballistic performance, are examined and analyzed as a function of composition and structure. The features of fracturing of these silicon carbide-based ceramics after ballistic impact are studied. The armor products (body armor plates with various configurations and flat tiles with a size of up to 300x300 mm for vehicular protection) are manufactured by slip casting or pressing. The armor systems based on studied SiC-based ceramics provide ballistic protection to Level III or Level IV, including satisfactory multi-hit performance, depending on the type and thickness of ceramics and backing materials.

INTRODUCTION

Ceramic armor systems consist, in general, of a monolithic ceramic body bonded with a soft but high tensile strength backing material such as special fiber lining (e.g. Kevlar^{TM}, Spectra^{TM} or fiberglass) and, sometimes, with soft metals (e.g. aluminum). Upon impact of the bullet with a high velocity (700-900 m/sec or greater) and kinetic energy of 2-4 kJ, the hard-faced ceramic is cracked and broken, and the residual energy is absorbed by the soft reinforced backing material. This backing material also supports post-impact fracturing of the ceramic body caused by the bullet and the bullet itself. Structure and properties of a ceramic facing material and its manufacturing features are significant factors affecting ballistic energy dissipation and, hence, performance of ballistic protection systems.

Among different structural ceramics, some types of oxide ceramics (mostly, alumina ceramics) and non-oxide ceramics (mostly carbides, nitrides, borides) are commonly used for ballistic protection. General properties of some armor ceramics are mentioned in the literature [1-9] and are summarized in Table 1.

Oxide ceramics, in particular alumina ceramics, have specific physical properties that are suited for armor applications. Although alumina ceramics have an elevated density (up to 3.95 g/cm^3), they are of low cost and may be manufactured by a variety of methods, i.e. slip casting, pressing and some others, without the use of expensive equipment, e.g. kilns with special controlled atmospheres. The firing temperature of this kind of ceramics is relatively low, and may range from 1500 to 1650°C depending on composition. Alumina-zirconia ceramics are also suitable for ballistic protection, but they are usually used when the weight of the armor material is not critical (the density of alumina-zirconia ceramics ranges from 4.2 to 4.5 g/cm^3 depending on composition).

In general, non-oxide armor ceramics such as boron carbide, silicon carbide, silicon nitride, aluminum nitride, titanium diboride and some others, including the materials based on their binary systems, have high physical properties and relatively low density (excepting titanium diboride-based ceramics) that are more beneficial for ballistic applications than alumina ceramics. However, these ceramics are usually manufactured by hot pressing that is relatively expensive and not very productive. Although pressureless sintered materials, such as commercially produced silicon carbide ceramics, are less expensive than hot-pressed materials, they are still relatively expensive because their manufacturing requires kilns with special controlled atmospheres and very high temperatures for sintering. Reaction-bonded silicon carbide (RBSC) and some other reaction-bonded carbide-based ceramics are considered as prospective materials for armor applications due to relatively lower cost than hot pressed or pressureless sintered ceramics, high physical properties and an ability to manufacture relatively large sized armor products [10].

Ceramic-matrix composites also demonstrate a high integrity after ballistic impact due to their mechanical properties and impact energy dissipation ability. The following ceramic-matrix composites are mentioned as armor materials [3]: ceramic reinforced with whiskers or fibers, such as compositions of Al_2O_3/SiC_w, Al_2O_3/SiC_f or Al_2O_3/C_f, and ceramics/particulate-based compositions (TiB_2/B_4C_p, TiB_2/SiC_p). Cermets such as Lanxide^{TM} composites based on silicon carbide infiltrated with aluminum, Ni/TiC, Al/B_4C_p and some others also demonstrate superior qualities. The majority of these materials are hot-pressed, and, therefore, expensive. Although some metal-infiltrated composites, such as Lanxide^{TM} SiC/Al composite, are not hot-pressed, they need special processes and equipment, they are also relatively expensive and are prone to difficult problems in manufacturing.

The development of lightweight and inexpensive ceramic armor materials is under ongoing consideration by both ceramic armor manufacturers and armor users. In this paper, pressureless sintered and reaction-bonded silicon carbide ceramics and the developed at Ceramic Protection Corporation (CPC) new ceramics in the initial systems based on $SiC-Al_2O_3$, $SiC-Si_3N_4-Al_2O_3$, $SiC-Si-Al_2O_3$ and $SiC-Si_3N_4-Si-Al_2O_3$ (which may also be considered as a composite due to their heterogeneous structure) are reviewed and studied. These ceramics have a relatively high level of physical properties, high ballistic impact energy dissipation ability and, as a result, a high level of ballistic performance. Different types of lightweight armor products with a variety of shapes may be manufactured without the use of expensive equipment.

EXPERIMENTAL
Materials and Manufacturing
The developed armor ceramics are based on formulations in the system of $SiC-Al_2O_3$ with, or without, small amounts of sintering aids. As a major raw material, black and green high-purity silicon carbide powders of different grades (SiC content is greater than 98-wt.%) commercially produced by Saint-Gobain Ceramics and Plastics (USA and Norway) are used. These powders have different average particle sizes and particle size distributions, e.g. the average particle sizes for the fine, medium and coarse powders are 1-3 μm, 25-35 μm, and 80-90 μm, respectively. Specially selected ratios of these powders are used in specific ceramic compositions. High-purity alumina raw materials commercially produced by Pechiney-Altech (France) and Alcoa World Chemicals (USA) have an Al_2O_3 content greater than 99.8-wt.% and an α-form content greater than 95-wt.%. The average median particle size and crystal size of the used alumina powders are 0.35-0.45 to 1.1-1.4 μm and their specific surface BET are 8-11 to 2.8-3.3 m^2/g, respectively. Silicon nitride powder commercially produced by Permascand AB – Akzo Nobel (Sweden) is used as an additional component for some developed compositions. This powder has a purity of greater than 99% with an $\alpha-Si_3N_4$ content of 93-95%; its average particle size and specific surface are 0.4-1.0 μm and 6-12 m^2/g, respectively.

Another group of the developed silicon carbide-based ceramics is based on the formulations in the systems $SiC-Si$, $SiC-Al_2O_3-Si$ and $SiC-Si_3N_4-Si-Al_2O_3$ with or without small amounts of sintering aids. Silicon carbide, silicon nitride and alumina raw materials used for these ceramics are described above. Silicon powders commercially produced by Permascand AB – Akzo Nobel (Sweden) have a purity of greater than 99%, their average particle size and specific surface are 35-45 and 6-8 μm and 0.2-0.6 and 2.0-2.2 m^2/g, respectively.

The batch compositions of the ceramics and the grades of raw materials used have been developed and selected to achieve an optimal particle size distribution and compaction, an optimal manufacturing ability and the occurrence of specific physical-chemical processes during firing conducted in special conditions.

Manufacturing processes of these new silicon carbide-based ceramics include the following steps:

- Water-based slip preparation using specifically selected dispersant and binder components. The slips have a solid content of 75-81-wt.% depending on the batch composition and a relatively low viscosity (Brookfield) of 85-120 cPs;
- Slip casting into plaster moulds providing required shapes;
- Drying the cast products;
- Firing the dried products. Kiln loading is specific for each product, and the firing conditions and the cycles have been specially optimized;
- Bonding of ceramics with a backing material; the ceramic surface preparation, as well as a selection of adhesive and thermal treatment of the glued ceramics with a backing material, have been optimized.

Ongoing laboratory and pilot-scale manufacturing studies have made it possible to produce armor products with a variety of shapes, e.g. tiles with a format of up to 300x300 mm and body armor plates with different shapes and sizes of 200-250x250-300 mm. Although only slip casting process was used in the present study, some other manufacturing methods (e.g. pressing, injection molding) also can be used.

Pressureless sintered silicon carbide (PLSSC) and reaction-bonded silicon carbide (RBSC) ceramics produced by other companies were manufactured based on their proprietary procedures. PLSSC ceramics were manufactured via pressing and slip casting technologies; RBSC ceramics were made by slip casting technology. The bonding procedures of the pressed tiles and slip cast armor plates and tiles made from these ceramics have been optimized in CPC.

Testing

Microstructure was studied using scanning electron microscopy (SEM). A selection of the test methods for evaluation of the studied heterogeneous silicon carbide-based ceramics is based on the structural features of this ceramics. The density of ceramics was tested using the water immersion method based on the Archimedes law. Sonic velocity and Young's modulus were tested by the ultrasonic technique measuring the longitudinal ultrasonic velocity in accordance with ASTM C769 and by the resonant frequency method in accordance with ASTM C885. The formula $E=V_l^2\rho(1+p)(1-2p)/(1-p)$ was used for the calculation, where E is the Young's modulus, V_l is the longitudinal sonic velocity measured in accordance with ASTM C769, ρ is the density, p is the Poisson's ratio. Three-point flexural strength was tested in accordance with ASTM C113. Impact strength was determined by measuring impact energy using a Charpy impact testing method with a swinging pendulum for rectangular bars. The specific energy (kJ/m^2) was calculated based on the measured impact energy and the sample cross-section dimensions. Vickers hardness of dense carbide-based ceramics was tested in accordance with ASTM C1327 at indentation loads of 0.1-1 kg. Fracture toughness (critical stress intensity factor) K_{lc} was determined using the indentation technique based on the samples prepared for Vickers hardness testing and was calculated using the well-known formula: $K_{Ic}=0.941Pc^{-3/2}$, where P is the indentation load and c is the crack length measured under microscope. Rockwell hardness was tested in accordance with ASTM E18 at loads of 150 kg (HRC testing) and 60 kg (HRA testing). These high loads have been selected due to general high hardness of silicon carbide. The test samples with required dimensions were cut from the actual products or from the test tiles with a format of 100x 100 mm or 150x150 mm produced by the mentioned technology.

The ballistic performance of the ceramics bonded with appropriate backing materials was tested in accordance with the NIJ 0101.03 and NIJ 0101.04 standards using weapons such as the M16, AK47 and some others (caliber 0.30). Depending on the application and the required level of protection, the ammunition 7.62x51-mm NATO Ball Full Metal Jacket (FMJ) with a lead core, 7.62x39-mm Russian Ball FMJ with a steel core, 7.62x63-mm armor-piercing APM2 FMJ with a tungsten carbide core and some others were used. Depending on the ammunition, the bullet weight, velocity and energy are varied, e.g. the bullet weight, velocity and energy for the mentioned projectiles are 9.65, 8 and 10.7 g, 830-900, 710-750 and 820-920 m/s, and approximately 3.5, 2.0 and 3.7 kJ, respectively. The bullet velocity during testing was measured using an optical chronograph. The trauma after shooting was measured in a Roma Plastilina modeling clay placed behind the armor system; the trauma in clay shows the transient deformation of the composite on the back of the system. The damage zone of the ceramics, including ceramic fragmentation, and the subsequent post-impact condition of the bullet, were observed. The flat tiles (100x100 mm or greater) with a thickness of 6-10 mm were used for single shot ballistic testing. Also the flat tiles (155x200 mm and 300x300 mm) with a thickness of 6-10 mm, as well as the actual plates with different configuration and the aforementioned thickness, were used for multi-hit ballistic testing with approximately 50-mm spacing between hits.

RESULTS AND DISCUSSION
Microstructure

The silicon carbide-based ceramics developed in the system SiC-(Si$_3$N$_4$)-Al$_2$O$_3$ have a heterogeneous structure (Fig. 1). It is formed by silicon carbide grains with different sizes ranging from a few microns to 120 μm bonded by a crystalline-glassy silicon carbide-aluminosilicate matrix. In the case of the use of silicon nitride constituent in the compositions, the bonding phase also includes silicon nitride and sialon. The compaction between silicon carbide grains is relatively high and it is achieved by specially selected ratios between the sizes of starting silicon carbide particles. A high level of the bonding between grains and a matrix is achieved due to a reaction-bonding mechanism. This reaction bonding occurs due to partial oxidation of some ingredients and the following high-temperature interaction of alumina with the products of the noted oxidation process. The structure in general, and the bonding phase in particular, are denser in the case of the nitride-based bonding phase. The use of small amounts of specially selected inorganic additives promotes the liquid phase formation and the interaction between phases during firing. These ceramics have some porosity, but the pore size is small (approximately several microns), and the pores are uniformly distributed. The majority of initial pores between grains, which occurred at the green body formation stage, disappeared during the firing process when the liquid phase is formed and the mullite crystals are initiated and grown. The surface of the ceramics has a higher content of a glassy phase than the middle. This is a result of the firing conditions and the features of the physical-chemical processes, defined by the diffusion mechanism, occurring during firing.

The ceramics in the systems of SiC-Si, SiC-Al$_2$O$_3$-Si and SiC-Si$_3$N$_4$-Al$_2$O$_3$-Si also have heterogeneous structures with the SiC grain sizes from a few microns to 120 μm but the bonding phase is represented by silicon nitride and sialon (in the case of the Al$_2$O$_3$-constituent) with a presence of a small amount of silicon (Fig. 2). The bonding phases in these ceramics are also formed due to reaction-bonding mechanisms occurred during nitridation. The ceramics also have small porosity with a very fine pore size. The sialon-bonded silicon carbide ceramics (i.e. the ceramics based on the initial system SiC-Si$_3$N$_4$-Al$_2$O$_3$-Si) have lower porosity comparatively with other materials from this group.

RBSC ceramics also have a heterogeneous microstructure formed SiC by grains of sizes up to 50-70 μm bonded by fine SiC crystals and residual silicon (approximately 10-12%) (Fig. 3). PLSSC ceramics have a dense homogeneous microstructure formed by uniform silicon carbide grains of a few microns size. These materials are dense and they have only residual close porosity.

Properties

The density of the studied silicon carbide-based ceramics from the system SiC-(Si$_3$N$_4$)-Al$_2$O$_3$ is relatively low, ranging from 2.7 to 3.2 g/cm^3 depending on the composition. The achievement of these low values is connected with some porosity of ceramics and, more importantly, with high-temperature chemical processes and, as a result, with a formation of new crystalline phase (mullite) with significantly lower density than starting alumina (3.2 vs. 3.95 g/cm^3). It is noted that the compositions containing silicon nitride constituent provide lower density and lower porosity among the studied ceramic materials. The ceramics have zero shrinkage or even slight expansion (the ratio between green and fired dimensions is in the range of 1.0-1.02 depending on compositions). The studied silicon carbide-silicon nitride and silicon carbide-sialon ceramics prepared via nitridation also have low density of 2.85-3.05 g/cm^3. These ceramics also have zero shrinkage connected with a formation of new phases due to reaction-bonding processes. RBSC and PLSSC ceramics have density of 3.05-3.1 g/cm^3 that correlates well with data known from literature.

The Vickers hardness test, traditionally used for evaluation of dense homogeneous ceramics, was used for evaluation of PLSSC ceramics. Vickers hardness testing was also used for the RBSC ceramics; however, due to their heterogeneous structure hardness was measured for the major grains and for microcrystalline matrix separately. Opposite to dense ceramics, Vickers hardness is not well-suited for slightly porous heterogeneous materials based on hard grains bonded by a matrix with lower hardness such as developed carbide-based ceramics. In the case of the Vickers hardness testing applied for these heterogeneous ceramics, the diamond pyramid-shaped indenter penetrates to the lower-hard matrix. For the studied heterogeneous ceramics, Rockwell hardness testing has been utilized. In this case, a spherical diamond indenter is applied to the surface of the ceramic, and the load is distributed on the sample surface more uniformly than in the case of the Vickers hardness testing.

Vickers hardness of the PLSSC ceramics is on a high level (HV0.3-HV1 values are in a range of 2200-2600 kg/mm^2). This dense ceramics demonstrates high values of Rockwell hardness (HRA is 90-92). Fracture toughness of this ceramics is relatively low (K_{Ic} values are 2.8-3.2 MPa.m$^{1/2}$). RBSC ceramics also demonstrated high hardness values. Due to their heterogeneous structure, Vickers hardness was measured for the primary silicon carbide grains and for the fine-crystalline matrix (HV1 values are in the ranges of 2350-2450 and 1600-1850 kg/mm^2, respectively). The lower hardness of a matrix may be a sequence of the presence of residual silicon and an incomplete formation of the silicon carbide phase. These ceramics have also high HRA values ranging from 90 to 92 units. A testing of HRC for these dense PLSSC and RBSC ceramics was not used due to their general high hardness and possible damage of the diamond indenter at 150-kg loadings. Fracture toughness values of the RBSC ceramics were also distinguished for the primary grains and for a matrix (2.4-2.9 and 3.4-4.3 MPa.m$^{1/2}$, respectively). The "level" of bonding in the RBSC ceramics affects mechanical properties of this ceramics. If the "level" of bonding is not high enough, i.e. if the high-temperature formation and development of the SiC-bonding phase is not completed and the amount of residual silicon is greater than 10-12%, hardness and some other properties of the matrix are lower. The Vickers hardness and fracture toughness testing of the fine-crystalline matrix should be considered as an important part of the characterization of RBSC ceramics. Some physical properties of dense silicon carbide ceramics are presented in Table 2.

Rockwell hardness of the studied ceramics developed in the starting systems of SiC-(Si$_3$N$_4$)-Al$_2$O$_3$ and SiC-(Si$_3$N$_4$)-(Al$_2$O$_3$)-Si depends on the ratio between relatively large silicon carbide grains and a crystalline-glassy matrix, on a compaction of the grains with different particle size (i.e. on the particle size distribution), and on "a level of bonding" between the grains and the matrix. The presence of microcracks between grains and the matrix connected with a possible stress occurrence during firing and cooling has an important effect on hardness values. The analysis of the Rockwell hardness test results showed that the samples with a higher content of relatively large silicon carbide grains coupled with a selected particle size distribution (where smaller particles occupy a space between larger particles), and with lower microcracks in the matrix, demonstrate higher hardness values. Considering materials from the group of SiC-Si, SiC-Al$_2$O$_3$-Si and SiC-Si$_3$N$_4$-Al$_2$O$_3$-Si, the ceramics made from the system SiC-Si$_3$N$_4$-Al$_2$O$_3$-Si, i.e. silicon carbide-sialon have a higher level of the bonding and, as a result, higher values of hardness.

Dense silicon carbide ceramics demonstrated high mechanical strength (flexural strength values are 350-400 and 240-280 MPa for PLSSC and RBSC, respectively). Mechanical strength of the studied heterogeneous silicon carbide-based ceramics is lower than fully sintered silicon carbide ceramics due to the presence of large SiC-grains in the structure of the materials and a residual porosity. Considering different compositions, flexural strength is higher if the ceramics have a finer microstructure and less coarse silicon carbide grains. However, the materials with relatively large particles, but with an optimized particle size distribution, have comparatively higher flexural strength. When considering these ceramics with different bonding phases, the sialon-bonded silicon carbide ceramics demonstrate higher values of flexural strength.

The impact energy transmission through materials and across boundaries occurs via a shockwave in ceramic armor systems. High sonic velocity of ceramics allows it to effectively remove energy from the impact zone. This is significant in considerations of armor, especially at high projectile velocity conditions. Dense silicon carbide ceramics demonstrated a high level of Young's modulus (380-420 and 315-370 GPa for pressureless sintered and RBSC ceramics, respectively). The developed silicon carbide-based ceramics possess a relatively high sonic velocity despite their heterogeneous structure and small porosity, comparable with sonic velocity achieved in some dense ceramics. They also have relatively high Young's modulus, achieving 280-310 GPa. Due to the demonstrated high sonic velocity for these ceramics it may be expected that their ability to dissipate impact energy also will be high.

Some physical properties for selected CPC-experimental compositions are presented in Table 3. These selected materials are distinguished by specific silicon carbide and alumina contents, by the presence and content of silicon nitride, by the particle size distribution of silicon carbide (i.e. by the ratio between silicon carbide constituents with different particle sizes), by the use of sintering aids and by the firing conditions. Considering the ceramics in the starting system of SiC-(Si$_3$N$_4$)-Al$_2$O$_3$, the materials have approximately the same level of mechanical properties, however, the ceramics with a Si$_3$N$_4$-constituent are more prospective for ballistic applications. Considering the another group, the sialon-bonded silicon carbide ceramics developed in the starting system of SiC-Si$_3$N$_4$-Al$_2$O$_3$-Si have significantly higher mechanical properties.

Ballistic Performance

The development and selection of ceramic armor is often based on a traditional approach, i.e. it is assumed that ceramic armor should be denser, harder and stronger (like ceramic cutting tools). This approach is correct in many cases, especially for the single-hit ballistic applications, when a ceramic tile has only to stop one projectile. In this case, hot-pressed or pressureless-sintered boron carbide

Ceramic Armor and Armor Systems

or titanium carbide or titanium diboride ceramics with greater values of hardness, mechanical strength and theoretical density present preferable options. However, practical experience has shown that dense ceramics with lower values of the aforementioned properties, such as some pressureless sintered silicon carbide-based ceramics and even alumina ceramics, can be successfully employed for some single-hit applications. For example, alumina ceramics with an Al_2O_3 content of 98-98.5-wt.% have successfully functioned for light-armor vehicle ballistic protection [8]. Contrary to data for dense homogeneous carbide-based ceramics, alumina ceramics with an Al_2O_3 content of 96-98-wt.% demonstrate formation of large chunks at the site of ballistic impact (although ceramic powdering is also observed), while the cracks are relatively short. The surrounding area of the impact zone is relative strong and, as practical experience shows, this kind of ceramics has been successfully employed for applications when protection from multi-hits are required [8]. In this case, microstructural features and an ability for ballistic impact energy dissipation have a higher importance than a high level of properties such as hardness, strength and some others.

However, ballistic test results showed that some dense carbide-based ceramics including boron carbide- and silicon carbide-based materials may demonstrate elevated ballistic shattering. Although these materials stop bullets thanks to their high hardness, and their trauma is relatively small, the damaged zone is characterized by many small ceramic micro-crack fractures and a comminuted powder. The surrounding zone with long cracks is not very strong and, as a result, these armor ceramics generally exhibit limited capabilities for multi-hit ballistic applications. This behavior of the dense homogeneous carbide-based ceramics under ballistic impact was confirmed by the testing with NATO Ball FMJ and APM2 projectiles for dense pressureless sintered silicon carbide ceramics. Despite the good bonding between ceramic tiles or plates and the backing material, the shattering was so intensive that the fractures (very small chunks and a comminuted powder) were hardly found in the system after shooting. Due to complete disintegration of this ceramics in the zone surrounding an impact area, this "zone" is not able to support the area of the next ballistic impact. This results in a limited capability of the PLSSC ceramics to withstand multi-hit situations. This weakening was observed especially after the third hit. High brittleness of this fine-crystalline ceramics (i.e. relatively high values of hardness and Young's modulus, but lower fracture toughness) may be one of the reason of such elevated fracturing and "powdering" under ballistic impact.

Ballistic test results for the newly developed SiC-Al_2O_3-(Si_3N_4)-based ceramics confirmed the importance of microstructure and ballistic impact energy dissipation ability on ballistic performance. Although these ceramics do not have extremely high physical properties comparable with some dense carbide-based ceramics, and even with alumina ceramics, they demonstrate a remarkable level of ballistic performance. As mentioned, this type of ceramics consists of silicon carbide grains bonded by a crystalline-glassy phase. The major constituent, i.e.

silicon carbide grains, possesses a high hardness value (Knoop hardness is 2700 or Moh's hardness is 9.2-9.5). The bonding matrix possesses a lower hardness value but it has sufficient mechanical properties to maintain the integrity of the material in ballistic applications. During ballistic impact the silicon carbide grains "stop" a high-velocity projectile (i.e. decrease its velocity significantly). The crystalline-glassy phase is fractured, but the propagated cracks stop at the silicon carbide grains surface or on the pore surface. The silicon nitride- and sialon-containing bonding phase reduces the crack propagation. The damaged zone is characterized by large and small chunks and powder. However, the balance of the ceramic plate is relatively strong, and it can provide further ballistic protection relatively close to the initial impact. The damage zone has a traditional conical shape with a locus of conoidal cracks initiated at the impact point and radial cracks initiated at the back surface. Other kinds of cracks (such as spall cracks) are also present. RBSC ceramics demonstrated a similar fracturing mechanism due to its heterogeneous structure and a relatively lower hard bonding phase. If the bonding matrix in the RBSC ceramics has lower hardness, i.e. if the "level" of bonding is not high enough, RBSC ceramics demonstrate lower ballistic performance. In this case, RBSC ceramics have elevated fracturing and elongated cracks at the ballistic impact and, as a result, greater trauma and other associated problems. It is observed, especially, at the multi-hit performance testing. The influence of the "level" of bonding, matrix properties and ballistic performance for armor ceramics, where "reaction-bonding" mechanism occurs, needs further studies. Some ballistic test results for the studied heterogeneous ceramics are illustrated in Fig. 4.

In general, the projectile energy is decreased by approximately three times after 20 μsec of the projectile impact [1], and a rate of the loss of energy and, therefore, ballistic performance, depend on the loss of a projectile mass and the loss of a projectile velocity. The loss of projectile mass occurs when the projectile is defeated by high-hardness ceramics and is eroded due to a high-friction effect. The coarse silicon carbide grains with a size of 50-120 μm used in the studied ceramic compositions have a great abrasiveness (e.g. greater than the smaller-size ones), and they erode the moving projectile significantly. For example, the observation of the projectiles (NATO Ball FMJ) after ballistic impact of the developed $SiC-Al_2O_3-(Si_3N_4)$-based ceramics with a high percentage of the coarse SiC-grains in the composition showed their highest damage and erosion among the other studied SiC-based ceramics. It is difficult to evaluate the decrease of a projectile velocity when it moves through a ceramic armor system with negative acceleration. However, it may be postulated that high-abrasive large-sized particles, such as the silicon carbide grains, promote remarkable friction between the moving projectile and the mentioned grains and, hence, promote the decrease of a projectile velocity greater than the fine-sized particles used for pressureless sintered ceramics.

The crack distribution in these heterogeneous ceramics is rather complex. In general, the energy dissipation through microcracking consists of a formation of numerous microcracks, characterized by the appearance of numerous stress micro-concentrators. It would appear that the energy from ballistic impacts is spent on the formation of numerous surfaces. If the microcracking occurs faster than general macrocrack propagation, the energy dissipation would be relatively more effective. Due to the presence of hard silicon carbide particles with different sizes and shapes, the direction in the crack propagation is re-oriented (comparatively with homogeneous microcrystalline ceramics) that could be considered as a positive factor in impact energy dissipation and, hence, in ballistic performance. Also the bridging of microcracks, which appears on the coarse silicon carbide particles, may also promote the energy dissipation and may decrease further crack propagation.

Due to the presence of large silicon carbide grains in these ceramic compositions, the compaction of the comminuted fragments and formed powder as a result of the projectile movement through this "coarse" ceramics appears to be less than for microcrystalline homogeneous ceramics. In this case, the penetration of the projectile through such a ceramic structure is more difficult. This compaction effect may also promote the achievement of high ballistic performance.

It has been demonstrated that armor systems based on the newly-developed silicon carbide-based ceramics bonded with appropriate aramid-based and/or aluminum backing materials are capable of defeating 7.62x39-mm and 7.62x51-mm Ball FMJ ammunition. Ceramics based on the starting $SiC-Al_2O_3-Si_3N_4$ system demonstrate higher ballistic performance (less trauma, less fracturing and shorter crack propagation) than other studied $SiC-Al_2O_3$ compositions. They provide ballistic protection to Level III or Level IV, depending on the ceramic and backing material thickness and armor system design (Level IV in conjunction with a Level II vest). These armor systems have satisfactory multi-hit ballistic performance (up to 6 or 9 hits depending on the projectile to one body-armor plate or tile) with acceptable levels of backface trauma, i.e. not greater than a 44-mm deformation in accordance with NIJ Standards. PLSSC ceramics with appropriate aramid-based backing are capable of defeating 7.62x39-mm and 7.62x51-mm Ball FMJ ammunition and 7.62x63-mm APM2 FMJ, but these ceramics are preferable for single-hit applications due to high brittleness. RBSC ceramics can defeat the mentioned projectiles and they demonstrated good multi-hit ballistic performance providing ballistic protection to Level III or Level IV.

CONCLUSIONS

A new type of light armor ceramics based on $SiC-Al_2O_3-(Si_3N_4)$ compositions has been developed and studied at CPC. These ceramics have a heterogeneous structure based on silicon carbide grains with a specially selected particle size distribution bonded by the aluminosilicate crystalline-glassy matrix (in some cases, with silicon nitride and sialon) or by silicon nitride-silicon carbide (in some cases, by sialon-silicon carbide) matrix formed by reaction-bonding mechanisms. These ceramics demonstrate remarkable physical properties, such as hardness, sonic velocity and some others and a high level of ballistic performance. High ballistic performance is mostly connected with the structural features of the ceramics and its ability to dissipate ballistic impact energy. Pressureless sintered and reaction-bonded silicon carbide ceramics also studied in the present work demonstrate high levels of physical properties depending on their compositions and microstructures and associated high ballistic performance. Heterogeneous RBSC and SiC-Si_3N_4-Al_2O_3 ceramics are also performed well in multi-hit applications.

REFERENCES

[1]. C.F. Cline, M.L. Wilkins, "The Importance of Material Properties in Ceramic Armor"; pp 13-18 in *DCIC Report 69-1*; Part I: "*Ceramic Armor*", 1969.

[2]. Soon-Kil Chung, "Fracture Characterization of Armor Ceramics", *American Ceramic Society Bulletin*, **69** [3] 358-66 (1990).

[3]. D.J. Viechnicki, M.J. Slavin, M.I. Kliman, "Development and Current Status of Armor Ceramics", *American Ceramic Society Bulletin*, **70** [6] 1035-39 (1991).

[4]. I.Yu. Kelina, Yu.I. Dobrinskii, "Efficiency of the Use of Silicon Nitride Ceramics as an Armor Material" (in Russian), *Refractories and Technical Ceramics*, [6] 9-12 (1997).

[5]. B. Matchen, "Application of Ceramics in Armor Products"; pp 333-342 in Key Engineering Materials, Vol. 122-124, *Advanced Ceramic Materials*. Edited by H. Mostaghasi. Trans. Tech. Publications, Switzerland, 1996.

[6]. R.G. O'Donnell, "An Investigation of the Fragmentation Behaviour of Impacted Ceramics", *Journal of Materials Science Letters*, **10**, 685-88 (1991).

[7]. V.C. Neshpor, G.P. Zaitsev, E.J. Dovgal, et al., "Armour Ceramics Ballistic Efficiency Evaluation"; pp 2395-401 in *Ceramics: Charting the Future*, Proceedings of the 8th CIMTEC (Florence, Italy, 28 June-4 July 1994). Edited by P. Vincenzini, Techna S.r.l., 1995.

[8]. E. Medvedovski, "Alumina Ceramics for Ballistic Protection", *American Ceramic Society Bulletin*, **81** [3] 27-32 (2002), [4] 45-50 (2002).

[9]. B.A. Galanov, O.N. Grigoriev, S.M. Ivanov, et al., "Structure and Properties of Shock-Resistant Ceramics Developed at the Institute for Problems in Materials Science"; pp 73-81 in *Ceramic Armor Materials by Design*, Ceramic Transactions, Vol. 134. Edited by J.W. McCauley, A. Crowson, W.A. Gooch, Jr., et al., 2002

[10]. M.K. Aghajanian, B.N. Morgan, J.R. Singh, et al., "A New Family of Reaction Bonded Ceramics for Armor Applications"; pp 527-539 in *Ceramic Armor Materials by Design*, Ceramic Transactions, Vol. 134. Edited by J.W. McCauley, A. Crowson, W.A. Gooch, Jr., et al., 2002

Table 1. Some Armor Ceramics and Their Properties

Ceramics	Density, g/cm^3	Vickers Hardness*, GPa	Fracture Toughness K_{Ic}, MPa.m$^{0.5}$	Young's Modulus, GPa	Sonic Velocity, km/sec	Flexural Strength, MPa
Alumina, Sintered	3.60-3.95	12-18	3.0-4.5	300-450	9.5-11.6	200-400
Alumina-Zirconia, Sintered	4.05-4.40	15-20	3.8-4.5	300-340	9.8-10.2	350-550
Silicon Carbide, Sintered	3.10-3.20	22-23	3.0-4.0	400-420	11.0-11.4	300-340
Silicon Carbide, Hot-pressed	3.25-3.28	20	5.0-5.5	440-450	11.2-12.0	500-730
Silicon Nitride, Hot-pressed	3.20-3.45	16-19	6.3-9.0	-	-	690-830
Boron Carbide, Hot-pressed	2.45-2.52	29-35	2.0-4.7	440-460	13.0-13.7	200-500
Titanium Diboride, Sintered	4.55	21-23	8.0	550	-	350
Titanium Diboride, Hot-pressed	4.48-4.51	22-25	6.7-6.95	550	11.0-11.3	270-700
Aluminum Nitride, Hot-pressed	3.20-3.26	12	2.5	280-330	-	300-400

*Tested at different loads

Table 2. Some Physical Properties of PLSSC and RBSC Ceramics

Property	PLSSC Ceramics	RBSC Ceramics
Density, g/cm^3	3.06 – 3.10	3.04 – 3.07
Rockwell Hardness HRA	91 - 93	90 – 92
Vickers Hardness HV1, kg/mm^2		
for major grains	2200 - 2600	2350 – 2450
for matrix		1600 – 1850
Fracture Toughness K_{Ic}, MPa.m$^{1/2}$		
for major grains	2.8 – 3.2	2.4 – 2.9
for matrix		3.4 – 4.3
Flexural Strength, MPa	350 - 400	240 – 280
Young's Modulus, GPa	400 - 420	370

Table 3. Some Physical Properties of the SiC-Based Heterogeneous Ceramics

Property	Ceramics in the group SiC-(Si$_3$N$_4$)-Al$_2$O$_3$	Ceramics in the group SiC-Si$_3$N$_4$-Al$_2$O$_3$-Si
Density, g/cm^3	2.7 – 3.2	2.85 – 3.05
Rockwell Hardness HRC	40 – 55	45 – 55
HRA	58 - 77	68 – 78
Flexural Strength, MPa	105 - 155	120 – 140
Impact Strength, kJ/m^2	1.86 – 2.24	-
Young's Modulus, GPa	240 - 310	260 – 280
Sonic Velocity, km/s	9.8 – 11.2	9.8 – 10.05

Fig.1. Microstructure of the ceramics in the SiC-Si$_3$N$_4$-Al$_2$O$_3$ system

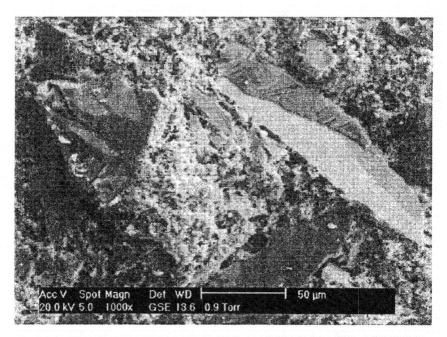

Fig. 2. Microstructure of the ceramics in the SiC-Si$_3$N$_4$-Al$_2$O$_3$-Si system

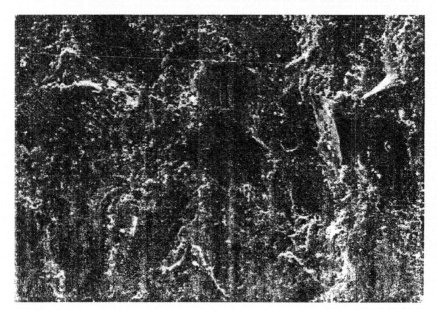

Fig.3. Microstructure of the RBSC ceramics

Ceramic Armor and Armor Systems

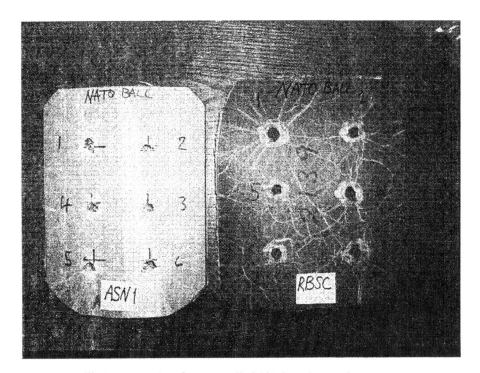

Fig. 4. Ballistic test results of some studied SiC-based ceramics

TOUGHNESS-HARDNESS TRADE-OFF IN ADVANCED SiC ARMOR

Marc Flinders, Darin Ray, and Raymond A. Cutler
Ceramatec, Inc.
2425 South 900 West
Salt Lake City, Utah 84119

ABSTRACT

Ceramic armor for advanced armor systems is envisioned to have both high hardness (HV10>28 GPa) and high fracture toughness (SEPB > 7 MPa-m$^{1/2}$). SiC candidate armor materials have been fabricated via solid-state or liquid-phase sintering to near theoretical densities. The toughness ranged between 2.5 and 8.5 MPa-m$^{1/2}$. Rietveld analysis was used to analyze polytypes and electron microscopy allowed microstructure to be correlated with hardness and toughness. Microstructures that give high toughness rely on crack bridging, which in turn is dependent on intergranular fracture. Hardness is lower for samples which fracture intergranularly and is not a strong function of grain size. Difficulties associated with making SiC both hard and tough simultaneously are discussed.

INTRODUCTION

Advanced SiC-based armor is desired such that the projectile is completely defeated without penetrating the ceramic armor. In order for this to occur, the ceramic must have both high hardness and high toughness[1]. A minimum short-crack toughness is predicted for a ceramic with a given yield strength in order to cause the projectile to dwell at the interface rather than penetrate the armor[1]. While this model has not yet been applied to liquid phase sintered (LPS) SiC, ceramics with a "soft" grain-boundary phase, they represent a class of ceramic armor where the toughness can vary from 2.5 to 8.5 MPa-m$^{1/2}$[2]. Ray, et al. showed that the general trend was for hardness to decrease as the long-crack toughness increased in a variety of SiC materials[2]. The purpose of this paper is to provide a correlation between microstructure and mechanical properties for these LPS SiC materials.

Initial LPS SiC materials were fine-grained and had only modest improvements in toughness, while retaining relatively high hardness[3]. Chia and Lau demonstrated that high toughness was possible in liquid phase sintered SiC and suggested that toughening occurred primarily due to microcracking between the YAG secondary phase and the SiC matrix[4]. Bocker and Hamminger[5] showed that elongated microstructures, similar to those produced in Si_3N_4-based ceramics, are possible by aging SiC sintered with a liquid phase. Padture[6] and Lee and Kim[7] noted that high toughness in SiC-based ceramics could be obtained by microstructural control. Toughness values above 5.5 MPa√m were achieved using this approach, which relied upon elongated grains that were most easily produced using β-SiC and allowing plate-like α-SiC to grow in a controlled manner. Seeding β-SiC with α-SiC promoted a duplex microstructure and aided in achieving high toughness. Their work showed that crack bridging aided the toughening process. Cao, et al.[8] increased the amount of Al previously used to sinter SiC and allowed an Al-B-C (ABC) additive system to promote SiC platelet formation. Similar to the work of Chia and Lau, their toughness showed an R-curve effect. Short-crack length toughness measurements, however, were still more than double the toughness of solid state sintered SiC[8].

Kim, et al.[9] demonstrated that α-SiC could be used to make tough LPS SiC provided that annealing allowed solution-precipitation to grow elongated α grains. Schwetz, et al.[10] have recently concluded that if high toughness is to be achieved with α-SiC starting powders, that low 4H polytypes are preferred in the starting powders so that the 6H to 4H transformation is maximized. They concluded that samples with higher 4H polytypes had higher toughness. Earlier work by researchers at Ford[11,12] demonstrated that Al additions promote 4H platelet growth, although they did not understand the ability of this growth to promote high toughness, as later shown by the work at Berkeley[8]. It is generally believed that β-SiC starting materials are more effective in forming elongated grains than α-SiC powders[13]. Seeding has recently been shown to have no effect on the β to α transformation rate and to lower the aspect ratio of SiC grains[14].

It is clear that there is still much confusion about what makes SiC tough. Little attention is paid to hardness or corrosion resistance, since these properties are degraded as compared to solid-state sintered SiC. In addition to showing the hardness-toughness trade-off, the present paper gives further insight into the role of microstructure in influencing room-temperature mechanical properties of SiC.

EXPERIMENTAL PROCEDURES

SiC-N, a state-of-the-art hot pressed armor material, was supplied by Cercom and Hexoloy SA, a commercially available solid-state sintered SiC, was purchased from St. Gobain. Processing of SiC powders with either the ABC or Y_2O_3-Al sintering additives has been reported previously[2]. Toughness was measured using the single-edge precracked beam (SEPB) technique[15] and hardness was measured using either Vickers[16] or Knoop indenters, with details as given previously[2]. Rietveld analysis[17-19] was used to determine SiC polytypes present in the densified samples with X-ray diffraction patterns collected from 30-80° 2-theta, with a step size of 0.02°/step and a counting time of 4 sec/step.

Polished samples of liquid phase sintered SiC were plasma-etched by evacuating and back-filling with 400 millitorr of CF_4-10% O_2 and etching for 20-40 minutes. Solid state sintered materials were etched in molten KOH at 550°C for 10-15 seconds. Grain size was determined by the line-intercept method, where the multiplication constant ranged between 1.5 (equiaxed grains) and 2.0 (elongated, plate-shaped grains) [20]. Typically, 200-300 grains were measured for each composition in order to get a mean grain size. The aspect ratio of the five most acicular grains in each of three micrographs was used to estimate a comparative aspect ratio.

RESULTS AND DISCUSSION

The densities of all samples were greater than 97% of theoretical and were generally greater than 99%. Rietveld analysis detected 3C, 4H, 6H, and 15R polytpes in the sintered samples. While toughness was well correlated with increasing alpha content in the samples, it was possible to make high toughness LPS SiC using either α or β SiC powders, as shown in Table 1, for samples hot pressed with a one hour hold time. This makes a correlation with starting beta SiC content meaningless. It is clearly shown by the data in Table 1 that the ABC system has the advantage of obtaining high toughness at a lower temperature than when YAG is used as a sintering aid, if similar amounts of liquid phase are used.

While the amount of 4H polytype in the annealed samples was somewhat correlated with fracture toughness when a single source of beta SiC powder was used to toughen SiC sintered with YAG, there was no correlation when a wide variety of LPS SiC samples were analyzed (see Figure 1). In contrast, grain size correlates relatively well with fracture toughness if samples that fail transgranularly are excluded (see Figure 2). Obviously, crack bridging can't occur if grains fracture transgranularly. If microcracking along grain boundaries were the primary toughening mechanism, one would not expect higher toughness with increasing grain size, since the number of interfaces decreases.

Table 1
Effect of Starting SiC Powder on Fracture Toughness

Description[2]	Sintering Temp.(°C)	SiC Starting Powder	Density (g/cc)	SEPB K_{Ic} (MPa-m$^{1/2}$)
SiC-3Al-2C-0.6B-BF17	1900	Beta	3.16±0.01	7.5±0.2
SiC-3Al-2C-0.6B-UF15	1900	Alpha	3.15±0.01	6.6±0.2
SiC-1Y-0.85Al-BF17	2200	Beta	3.24±0.01	7.8±0.3
SiC-1Y-0.85Al-UF15	2200	Alpha	3.25±0.01	6.6±0.1
SiC-2Y-1.7Al-BF17	2200	Beta	3.24±0.01	6.9±0.4
SiC-1Y-1.7Al-UF15	2200	Alpha	3.25±0.01	6.7±0.5

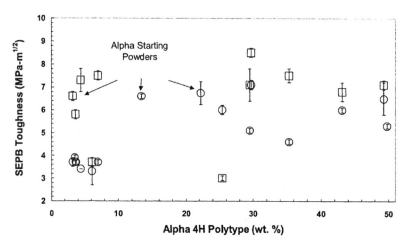

Figure 1. Fracture toughness as a function of 4H polytypes in the sintered SiC. Circles are SiC densified with YAG and squares are ABC SiC. The amount of 4H polytype in sintered SiC does not correlate with toughness .

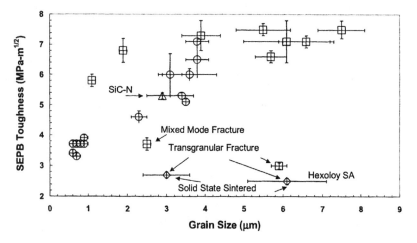

Figure 2. Fracture toughness as a function of grain size. Grain size correlates fairly well with fracture toughness if only samples that fail intergranularly are considered. Circles are SiC with YAG, squares are ABC SiC, the triangle is SiC-N and diamonds are solid state sintered SiC. Crack bridging aids in toughening LPS SiC.

Ceramic Armor and Armor Systems

When all the data are considered, the toughness correlates best with grain size. The exception to this correlation is SiC where fracture is transgranular either due to the lack of a grain boundary phase (solid state SiC sintered with B and C additions) or samples where the grain boundary chemistry allows transgranular fracture (see Figures 3 and 4). This is very apparent in the case of ABC SiC, where decreasing the Al from 3 wt. % to 1.5 wt. % changes the fracture mode from primarily intergranular to mostly transgranular fracture (see Figure 4). When solid-state sintered SiC is compared with liquid phase sintered SiC at the same grain size, it is apparent that intergranular fracture aids in toughness.

Figure 5 shows that the increasing aspect ratio is the main reason why grain size correlates with fracture toughness. With more accurate aspect ratio data it is possible that it would correlate better with toughness than grain size. Figure 6, however, shows that while high aspect ratio grains can help a fine-grained ABC SiC have high toughness, there is not a one-to-one correlation of aspect ratio with toughness.

In summary, the toughness of LPS SiC is controlled by grain boundary chemistry, which allows intergranular fracture and subsequent grain bridging resulting in high toughness for long, sharp cracks. The starting SiC polytypes or the resulting polytypes have little influence on toughness. Relatively tough (5-6

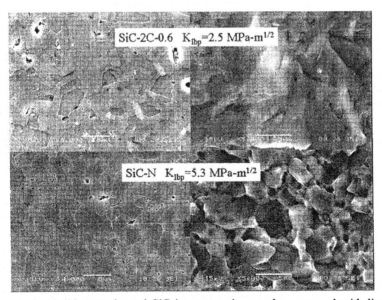

Figure 3. Solid state sintered SiC in upper micrographs compared with liquid phase sintered SiC-N of similar grain size. Note how intergranular fracture promotes higher toughness. All markers are 5 μm long.

Figure 4. Grain boundary chemistry is important in influencing the fracture toughness of LPS SiC. Transgranular fracture causes low toughness. All markers are 5 μm long.

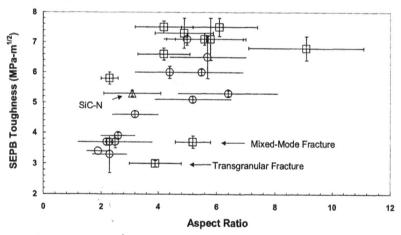

Figure 5. Fracture toughness of LPS SiC as a function of SiC aspect ratio, showing a general increase in toughness with increasing aspect ratio for samples that fail intergranularly. Circles are SiC sintered with YAG, squares are ABC SiC, and the triangle is SiC-N.

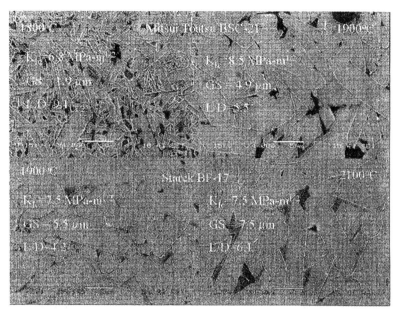

Figure 6. Microstructures of LPS SiC showing grain evolution with increasing temperature. Markers are 5 μm long. High fracture toughness is obtained by having elongated grains that can bridge cracks. Higher aspect ratios do not necessarily result in higher toughness.

MPa-m$^{1/2}$) SiC can be made with modest aspect ratios (2-3) by controlling the grain boundary chemistry, which allows intergranular fracture to occur.

Previous work showed that hardness measurements are problematic for these materials due to intergranular cracking making measurements difficult and the load dependent upon hardness[2]. In spite of these difficulties it is possible to assess the hardness-toughness relationship, as shown in Figure 7(a). When Knoop hardness is measured at a load of one kilogram, the trend is not obvious for SiC sintered with YAG (see Figure 7(b)). What is very apparent, however, is that solid state sintered SiC with low porosity has higher hardness than LPS SiC. It is also very obvious that the choice of liquid phase governs the hardness, with YAG giving higher hardness than ABC additives at similar volume fractions.

Figure 8 clearly shows that grain size does not control hardness in LPS SiC, contrary to that observed in many other ceramics[21]. Hardness in LPS SiC is controlled primarily by the choice of the grain boundary chemistry. This suggests that intergranular fracture allows grains to move more readily under the indentation load, resulting in lower hardness with increasing loads[2]. It is very clear that grain size is not well correlated with hardness in the same manner that it affects toughness.

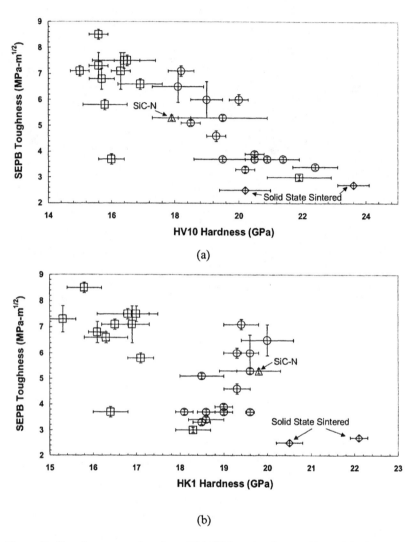

(a)

(b)

Figure 7. Toughness as a function of (a) Vickers hardness with a 10 kg load, or (b) Knoop hardness with a 1 kg load. As SiC increases in toughness, it generally loses hardness resulting in a hardness-toughness trade-off. ABC SiC (square symbols) is softer than SiC sintered with YAG (circles) as a secondary phase at the same toughness. The hardness results are dependent on the method used to measure hardness.

Ceramic Armor and Armor Systems

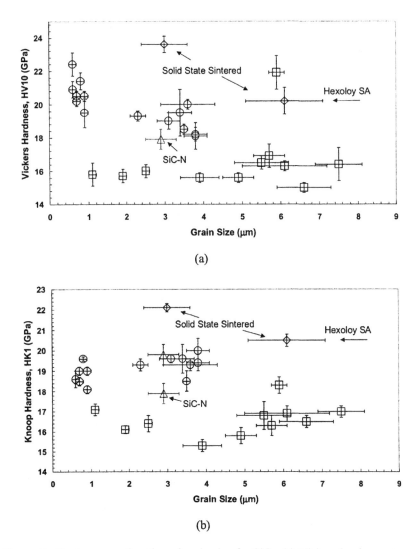

Figure 8. Hardness as a function of grain size for SiC. (a) Vickers hardness at a load of 10 kg, and (b) Knoop hardness at a load of 1 kg. Circles are SiC sintered with YAG, squares are ABC SiC, triangle is SiC-N, and diamonds are solid-state sintered SiC. Note how hardness is more dependent on the choice of the sintering additive than the grain size. LPS SiC is softer than solid state sintered SiC.

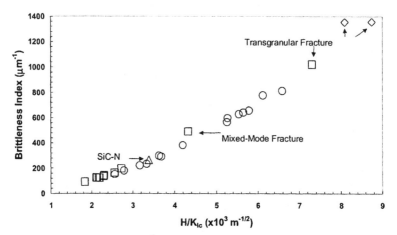

Figure 9. Brittleness (HE/K_{Ic}^2) as a function of H/K_{Ic}. Transgranular fracture leads to high brittleness. Tough LPS SiC compositions have a brittleness index similar to Si_3N_4 materials[22].

Quinn and Quinn have suggested that B, the brittleness index, which is equal to HE/K_{Ic}^2, where H is load-independent Vickers hardness and E is elastic modulus, may have predictive capability in ceramic design[22]. Despite the fact that we do not have load-independent hardness[2], it is still instructive to view our data relative to what they reported. Figure 9 shows the brittleness index as a function of H/K_{Ic}, which is simply another parameter for assessing brittleness in ceramics[22]. For reference purposes, B=150-250 μm^{-1} for Si_3N_4, 450 μm^{-1} for Al_2O_3, and 1100-1325 μm^{-1} for solid-state sintered SiC[22]. The agreement between solid-state sintered SiC in the present work and that same material analyzed by Quinn and Quinn is good, suggesting that many of the LPS SiC materials are similar in ductility to Si_3N_4.

It is demonstrated that it is easy to make SiC tough but it is problematic to make materials which are both hard and tough at the same time. In order to make materials hard one needs to eliminate the grain boundary phase, but to make SiC tough, it is imperative to have a grain boundary phase to get intergranular fracture and crack bridging. This apparent dichotomy makes the attainment of hardness and toughness in monolithic SiC a challenging problem.

What is lacking in the present investigation is the influence of room temperature mechanical properties on ballistic performance. Ballistic testing is needed to elucidate the role of the hardness-toughness trade-off on penetration characteristics. In order to achieve dwell and projectile defeat, based on the LaSalvia model[1], one needs increase in toughness without decreasing hardness. This has yet to be demonstrated for SiC ceramics due to the inherent hardness-toughness trade-off.

CONCLUSIONS

Rietveld analysis was effective in showing that the polytypes formed in a wide variety of liquid phase sintered SiC materials did not control fracture toughness. The SEPB fracture toughness was dependent on the grain size (i.e., aspect ratio) for samples which fractured intergranularly. High toughness (>6.5 MPa-m$^{1/2}$) could be made using either alpha or beta starting powders. The grain boundary chemistry has a strong influence on the toughness and hardness of SiC.

Hardness for LPS SiC, which is dependent on both load and test technique, is not strongly influenced by the grain size or aspect ratio, but is affected by the grain boundary chemistry. Samples with YAG as a secondary phase are harder than ABC SiC at the same level of toughness with similar amounts of sintering additives. It is difficult to make SiC both hard and tough at the same time because intergranular fracture is necessary for toughening, but weak grain boundaries allow easier displacement of grains during hardness testing. Ballistic test data are needed to determine if any correlation exists between ballistic performance and room-temperature brittleness.

ACKNOWLEDGEMENT

This SBIR research and development project was performed for the U.S. Army under contract DAAD17-02-C-0052.

REFERENCES

1. J. C. LaSalvia, "A Physically-Based Model for the Effect of Microstructure and Mechanical Properties on Ballistic Performance," *Ceram. Sci. and Eng. Proceedings*, 23[3], 213-20 (2002).
2. D. Ray, M. Flinders, A. Anderson, and R. A. Cutler, "Hardness/Toughness Relationship for SiC Armor," to appear in *Ceram. Sci. and Eng. Proceedings*, **24** (2003).
3. R. A. Cutler and T. B. Jackson, 'Liquid Phase Sintered Silicon Carbide," pp. 309-318A in *Third International Symposium on Ceramic Materials and Components for Engines*," ed. by V. J. Tennery (Am. Ceram. Soc., Westerville, OH. 1989).
4. K. Y Chia and S. K. Lau, "High-Toughness Silicon Carbide," *Ceram. Eng. Sci. Proc.*, **12**, 1845-61 (1991).
5. W. Bocker and R. Hamminger, "Advancements in Sintering of Covalent High-Performance Ceramics," Interceram, 40[7] 520-25 (1992).
6. N. P. Padture, "In Situ-Toughened Silicon Carbide," *J. Am. Ceram. Soc.*, 77[2] 519-23 (1994).
7. S. K. Lee and C. H. Kim, "Effects of α-SiC versus β-SiC Starting Powders on Microstructure and Fracture Toughness of SiC Sintered with Al_2O_3-Y_2O_3 Additives," *J. Am. Ceram. Soc.*, 77[6] 1655-58 (1994).
8. J. J. Cao, W. J. MoberlyChan, L. C. DeJonghe, C. J. Gilbert, and R. O. Ritchie, "In Situ Toughened Silicon Carbide with Al-B-C Additions," *J. Am. Ceram. Soc.*, 79[2] 461-69 (1996).

9. J. Y. Kim, Y. W. Kim, M. Mitomo, G. D. Zhan, and J. G. Lee, "Microstructure and Mechanical Properties of α-Silicon Carbide Sintered with Yttrium-Aluminum Garnet and Silica," *J. Am. Ceram. Soc.*, **82**[2] 441-44 (1999).

10. K. A. Schwetz, E. Schäfer, and R. Telle, "Influence of Powder Properties on In-Situ Platelet Reinforcement of LPS-SiC," *DKG*, **80**[1-2] E1-E6 (2003).

11. R. M. Williams, B. N. Juterbock, S. S. Shinozaki, C. R. Peters, and T. J. Whalen, "Effects of Sintering Temperatures on the Physical and Crystallographic Properties of β-SiC," *Am. Ceram. Soc. Bull.*, **64**[10] 1385-89 (1985).

12. S. S. Shinozaki, R. M. Williams, B. N. Juterbock, W. T. Donlon, J. Hangas, and C. R. Peters, "Microstructural Developments in Pressureless-Sintered β-SiC Materials with Al, B, and C Additions" *Am. Ceram. Soc. Bull.*, **64**[10] 1390-93 (1985).

13. H. Xu, T. Bhatia, S. A. Deshpande, N. P. Padture, A. L. Ortiz, and F. L. Cumbrera, "Microstructural Evolution in Liquid-Phase-Sintered SiC: Part I, Effect of Starting Powder," *J. Am. Ceram. Soc.*, **84**[7] 1578-84 (2001).

14. S. A. Deshpande, T. Bhatia, H. Xu, N. P. Padture, A. L. Ortiz, and F. L. Cumbrera, "Microstructural Evolution in Liquid-Phase-Sintered SiC: Part II, Effects of Planar Defects and Seeds in the Starting Powders," *J. Am. Ceram. Soc.*, **84**[7] 1585-90 (2001).

15. ASTM C 1421-99, Standard Test Methods for Determination of Fracture Toughness of Advanced Ceramics at Ambient Temperature, pp. 641-672 in *1999 Annual Book of Standards* (ASTM, Philadelphia, PA 1999).

16. ASTM C 1427-99, Standard Test Methods for Vickers Indentation Hardness of Advanced Ceramics, pp. 480-487 in 1999 *Annual Book of Standards* (ASTM, Philadelphia, PA 1999).

17. H. M. Rietveld, "A Profile Refinement Method in Neutron and Magnetic Structures," *J. Appl. Crystallogr.*, **2**, 65-71 (1969).

18. D. L. Bish and S. A. Howard, "Quantitative Phase Analysis Using the Rietveld Method," *J. Appl. Crystallogr.*, **21**, 86-91 (1988).

19. A. L. Ortiz, F. L. Cumbrera, F. Sanchez-Bajo, F. Guiberteau, H. Xu, and N. P. Padture, "Quantitative Phase-Composition Analysis of Liquid-Phase-Sintered Silicon Carbide Using the Rietveld Method," *J. Am. Ceram. Soc.*, **83**[9] 2283-86 (2000).

20. E. E. Underwood, *Quantitative Stereology*, (Addison-Wesley, Reading, MA. 1970).

21. A. Krell, "A New Look at the Influences of Load, Grain Size, and Grain Boundaries on the Room Temperature Hardness of Ceramics," *Int. J. Refrac. Mater.*, **16**, 331-35 (1998).

22. J. B. Quinn and G. D. Quinn, "Indentation Brittleness of Ceramics: A Fresh Approach," *J. Mater. Sci.*, **32**, 4331-46 (1997).

DEVELOPMENT OF A PRESSURELESS SINTERED SILICON CARBIDE MONOLITH AND SPECIAL-SHAPED SILICON CARBIDE WHISKER REINFORCED SILICON CARBIDE MATRIX COMPOSITE FOR LIGHTWEIGHT ARMOR APPLICATION

T.M. Lillo
Bechtel BWXT, Idaho
P.O. Box 1625
Idaho Falls, ID 83415-2218

D.W. Bailey
Superior Graphite Co
4059 Calvin Drive
Hopkinsville, KY 42240

D. A. Laughton
Superior Graphite Co
10 S. Riverside Plaza
Chicago, IL 60606

H.S. Chu
Bechtel BWXT, Idaho
P.O. Box 1625
Idaho Falls, ID 83415-0325

W. M. Harrison
Superior Graphite Co
4059 Calvin Drive
Hopkinsville, KY 42240

ABSTRACT

Thick tiles were fabricated to address higher ballistic threats in a continuing effort to develop a pressureless sintered α-silicon carbide ceramic for lightweight armor applications. Tiles with a thickness of 1.9 cm and 2.8 cm were pressed and sintered from a spray-dried SiC powder and evaluated for density and mechanical properties. These thick tiles exhibited a gradient in density and mechanical properties at the center of the tile with larger gradients present in the 2.8 cm thick tile. The Modulus of Rupture ranged from a low of approximately 300 MPa to a maximum of 500 MPa. It is concluded that adjustments to green part fabrication methods and/or to the sintering schedule are required to eliminate gradients in density and mechanical properties. Additionally, composites utilizing shape-modified, silicon carbide ceramic whiskers are investigated with the goal of increasing the toughness of an otherwise brittle silicon carbide ceramic. The ceramic reinforcement whiskers are heat treated at high temperature in an inert atmosphere to produce rounded ends (bone-shaped whiskers) via atomic transport. The whisker-reinforced composite exhibited reduced sintered density and exaggerated grain growth. Physical/mechanical properties, photomicrographs and ballistic test results are reported and compared to commercial, armor-grade silicon carbide.

INTRODUCTION

A pressureless sintered SiC ceramic was reported at the 27th Annual Cocoa Beach Conference and Exposition on Advanced Ceramics and Composites that exhibited slightly lower mechanical properties than hot pressed SiC ceramics[1]. However, this material performed similar to hot pressed material during ballistic tests. Finer starting particle sizes, high green density and relatively short sintering times were used to develop microstructures that more closely resemble those of hot pressed ceramics. One goal of this work is to use these strategies to develop fine-grained silicon carbide in thicker tiles to address higher ballistic threat levels.

Even though a high-density, fine-grained silicon carbide ceramic is capable of defeating ballistic projectiles it is not typically capable of withstanding more than one direct hit due to its brittle nature. Historically additions of whiskers have resulted in minimal improvement, and in some cases, degradation in fracture toughness[2,3]. It is thought that the lack of improvement is due to lower green density brought on by inefficient packing of the whiskers and/or the sharp corners at the whiskers ends produce stress concentrations that result in lower than expected fracture toughness. The second part of this work reports on efforts to incorporate shape-modified silicon carbide whiskers into the pressureless sintered SiC to increase fracture toughness.

MANUFACTURING

The starting powder used to fashion 10.2 cm x 10.2 cm ceramic armor tiles was Superior Graphite's HSC490 grade submicron alpha silicon carbide. A slurry was made using this powder and various proprietary binders, die lubricants and sintering aids. The organic binders and lubricants were added in a ratio that comprised from 2-10% of the total solids weight, with boron carbide used as the sintering aid. The slurry was flash spray-dried yielding a rheological product. The spray-dried powder was pressed to form green bodies with a density between 1.75-2.10 g/cc, then dried overnight in a standard drying oven. The resultant green bodies were then pressureless sintered in a furnace with a cycle time that did not exceed 24 hours. The maximum temperature fell between 2050°C to 2175°C. These tiles were then machined to meet dimensional specifications. Tiles with a thickness of 12.7 mm, 19.1 mm and 27.9 mm were made.

Surface diffusion and/or vapor phase atomic transport at elevated temperature was utilized to produce the rounded ends on the whiskers (initially 1.5 μm diameter by 18 μm long, Alfa Aesar). The whiskers were loaded loosely into an uncovered graphite crucible. The crucible was heated to 700°C and held for 2 hours in a flowing helium-2% methane atmosphere. Since decomposition of methane into hydrogen and free carbon is almost complete around 500°C[4], carbon would deposit on the whisker surfaces, helping prevent sintering together of the whiskers during the higher temperature shape modification step. After the carbon deposition step the crucible was heat further to 1700°C and held for an additional 2 hours under high purity argon. The modified whiskers were then wet mixed with the spray-dried HSC490 powder using Darvan C as a dispersant. The slurry was roll mixed in a polyethylene bottle containing 6.35 mm diameter SiC balls. After mixing the slurry was poured into a dish and allowed to dry. After crushing, the powder was dry pressed in a 76.2 mm diameter die at 34.5 MPa and then cold isostatic pressed at 414 MPa. Samples were then sintered in the same manner as described above for the monolithic SiC.

All samples were evaluated for density using the Archimedes principle and microstructure using optical metallography. Additionally, tiles without SiC fibers were evaluated for Modulus of Rupture (MOR) with 4-pt bend bars prepared to ASTM standard C1161-94 (bend bar dimensions – 3.0 mm x 4.0 mm x 50 mm). Bend bars were made from regions near the edge and regions in the center of the tiles to assess variations in MOR values across the tiles. Multiple bars were made at each region and their position through the thickness of the tile was also tracked to assess MOR variations through the tile thickness. Following MOR determination the bend bars were evaluated for density to track variations of density across and through the thickness of the 19.1 and the 27.9 mm thick tiles. The 12.7 mm thick tile was too thin to evaluate MOR and density as a function of the tile thickness.

RESULTS
Pressureless Sintering of Monolithic Silicon Carbide

Table I summarizes the physical and mechanical properties exhibited by both the pressureless sintered samples of this study as well as for commercially available SiC material. Simple press and sinter of the HSC490 sub-μm α-SiC powder produced densities typically exceeding 98% of theoretical, (>3.16 g/cm^3), with virtually no warpage or edge cracking. For comparison purposes, Cercom SiC-B, produced by hot pressing, exhibited essentially full density at 99.9% of theoretical. (A post-sintering hot isostatic pressing of a 12.7 mm HSC490 tile at 207 MPa and 1800°C for 2 hours only increased the density marginally to 98.6% of theoretical from 98.2% before HIP processing.)

The average MOR was evaluated from 24, 20 and 20 four-point bend bars taken from the 12.7, 19.1 and the 27.9 mm thick HSC490 tiles, respectively. In each case bend bars were cut from a single tile. The average MOR of all the HSC490 tiles is below the value given for Cercom SiC-B. The Weibull parameters were also calculated from these MOR data and the Weibull modulus and characteristic strength are, again, low compared to the values given for commercially hot pressed SiC, Table 1.

Table I. Mechanical Properties of Pressureless-sintered and Commercial SiC

Typical Properties	Cercom		Superior Graphite/INEEL			
Grade of SiC	SiC B	SiC N	HSC490 12.7 mm	HSC490 19.1 mm	HSC490 27.9 mm	HSC490 /Whiskers
Bulk Density g/cm3	3.20	3.20	3.17	3.18	3.17	3.05
Average Grain Size μm	3-5	3-5	5-10	5-12	5-15	-
Average Flexural Strength, MPa (4-Pt MOR)	560	580	286	382	421	-
Characteristic Strength, MPa	595	600	307	404	455	-
Weibull Modulus (m)	11	17	6.5	8.5	5.9	-

Table I also shows the average flexural strength increasing as thickness increases for the HSC490 material. The reason for this trend is revealed in Figure 1 where the MOR and density as a function of location in the tile is given. Figure

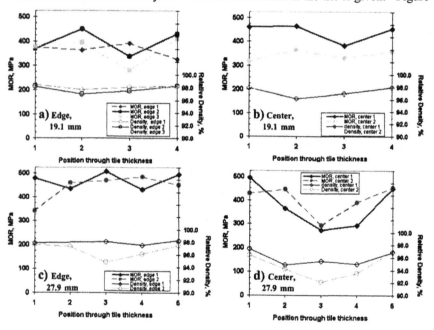

FIGURE 1. Plots of MOR and density at the edge and in the center through the thickness in the 19.1 mm (a & b) and 27.9 mm thick tiles (c & d), respectively.

1a (tile edge) and 1b (tile center) show the MOR and density are approximately constant through the thickness of the tile with possibly a slight decrease in density in the middle of the tile. However, Figure 1d (tile center) reveals that the 27.9 mm thick tile has a lower density region in the center of the tile and a corresponding decrease in the value for the MOR. These observations would seem to indicate the MOR and density are a function of tile thickness and mostly likely develop during either cold pressing or sintering. Dry pressing of thick parts can lead to gradients and green density that are carried through sintering[5]. Temperature gradients in the tile during firing also may become significant as the tile thickness increases due to longer thermal diffusion distances. Thick tiles will be under-sintered with relatively low density in the center of the tile if the sintering schedule is not adjusted appropriately. Figure 2 was generated to gain a more detailed understanding of the MOR and density values in Figure 1. The

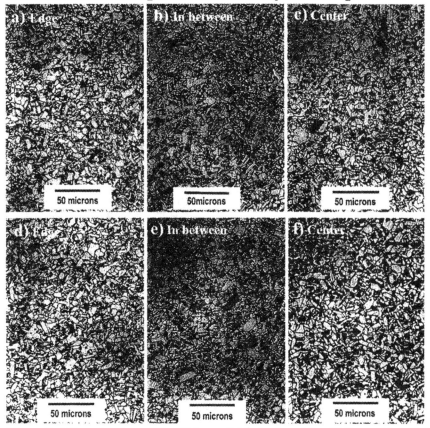

Figure 2. Microstructure of the 19.1 mm thick (a-c) and the 27.9 mm thick (d-f) tiles at regions near the edge (left photo), in between the center and edge (middle) and at the center (right) of the tiles.

microstructure in 19.1 mm thick tile varies very little from the edge (Fig. 2a) to the center (Figure 2c) of the tile while the 27.9 mm thick tile exhibits a slightly increasing grain size and porosity from the edge (Fig. 2d) to the center (Fig. 2f) of this tile. It is also noted that the grain size in Fig. 2f is slightly larger than the corresponding location in the thinner tile (Fig. 2c). The microstructures in Figure 2 are in agreement with the variations in MOR and density shown in Fig. 1.

In general, the sintering conditions produced a uniform grain size and density through out the 19.1 mm thick tile. These same sintering conditions resulted in a dense outer edge and a relatively low density interior in the 27.9 mm thick tile. Thickness-dependent modifications to the sintering schedule are expected to produce uniform material with MOR values approaching or exceeding 500 MPa.

The relatively low MOR values for the 12.7 mm thick tile shown in Table 1 are somewhat puzzling. The grain size and density appear to be equivalent to the 19.1 mm thick tile that exhibits greater values for the MOR. Possible explanations include the effects of a slightly different sintering schedule (again, proprietary) experienced by this tile as well as the fact that these bend bars were manufactured "in-house" while the bend bars for the 19.1 and the 27.9 mm thick were made by a commercial vendor, experienced in bend bar fabrication. The variation of MOR from the edge to the center of this tile as well as a TEM micrograph of the microstructure is shown in Figure 3. Both the density and the MOR are approximately constant across the tile. The microstructure in Fig. 3b

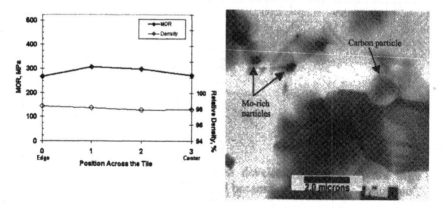

Figure 3. Properties and microstructure of the 12.7 mm thick SiC tile.

shows the presence of Mo-rich particles at triple junctions as well as carbon particles. Both types of particles were identified by energy dispersive x-ray spectroscopy in the TEM. The Mo-rich particles are relatively small at ~0.5 μm while the carbon particles are on the order of 1-1.5 μm in diameter. The carbon particles may adversely affect the MOR in this tile as well as the other tiles if they are present. Steps are being taken to eliminate the Mo-contamination during powder production and milling and to decrease the presence of excess carbon.

Whisker Shape Modification

Figure 4 shows examples of SiC whiskers that have been heat treated to modify their shape. Refer to reference [1] for a more complete description of the shape modification process. The proper amount of "spherodization" to yield the optimum improvement in fracture toughness of a composite is not known at this time and this work merely demonstrated that it can produce a shape change in the whiskers. Optimization of the level of spherodization was not explored at this time. Figure 4 shows that 2 hours at 1700°C is sufficient to produce rounded ends on the whiskers.

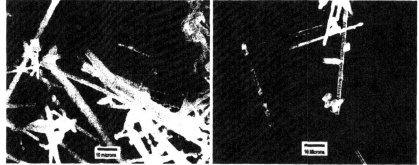

Figure 4. a) As-received SiC whiskers and b) after heat treatment, 1700°C, 2 hrs.

The incorporation of the shape-modified whiskers into the 490DP matrix at a level of 5% yielded disappointing results. In addition to poor sintered density, Table 1, highly elongated grains are seen to have evolved, Fig. 5. These grains are more extensive than the whiskers that were added and therefore are expected to mask any benefit the shape-modified whiskers may have imparted. This type of microstructure is

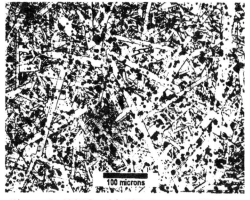

Figure 5. 490DP with 5% shape-modified whiskers.

usually attributed to the transformation of β-SiC to α-SiC during high temperature sintering[6]. This is somewhat unexpected in these samples since the 490DP powder used for the matrix is α-SiC already. The shape-modified whiskers are β-SiC and it was expected that only the whiskers would transform to α-SiC during high temperature sintering. In future work it will first be necessary to address the reason for the poor sintered density and then delve into the development of the elongated grains.

BALLISTIC TESTS AND RESULTS

At this time the only materials subjected to ballistic testing are the 12.7 mm thick 490DP, 490DP with shape-modified whiskers (5% by volume) additions and a commercial hot-pressed SiC material (Cercom, SiC-B) which was used as a benchmark. These materials were tested initially using a Depth-of-Penetration (DOP) test. (For more details on the DOP test and test conditions refer to [1].) The mass efficiency for these materials is presented in Table II. As shown the mass efficiency of the pressureless sintered monolithic SiC is ~20% lower than that of the SiC-B benchmark material. There is no further improvement in ballistic strength with the post-hot-isostatic press process, presumably due to the insignificant increase in density. As expected the SiC whisker-reinforced SiC composite performed poorly due to insufficient densification and grain growth during sintering.

Table II. DOP with Mass Efficiency Results

Target ID	Process	Density, g/cc	Thickness, cm	Velocity, m/sec	DOP, cm	Mass Efficiency
502-SP-4	HSC490, pressureless sintered	3.17	0.45	861	0.335	5.7
502-SP-2	HSC490, pressureless sintered, HIP	3.18	0.43	844	0.366	5.3
5F0561B-2	HSC490 + 5% whiskers, pressureless sintered	3.05	0.64	837	0.757	3.8
Cercom SiC-B	Pressure assisted densification (PAD)	3.22	0.44	847	0.147	7.0

Ballistic limit tests were also performed on the material configured in a structural armor system to further verify the ballistic performance. PAD SiC and Si₃N₄ materials from Ceradyne and PAD SiC-B materials from Cercom were used as benchmark materials. The ballistic limit tests were performed by the Armor Mechanics Branch of the Army Research Laboratory, Aberdeen Proving Ground, MD. All test materials have an initial thickness of 12.7 mm and they were

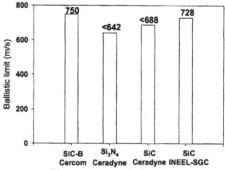

Figure 6. Ballistic limit test results
Note: Valid V50 for Ceradyne material was not determined. Numbers given are high-end estimates based on number of samples available.

Ceramic Armor and Armor Systems

adhesively bonded to the face of (152 x 152 x 25.4) mm^3 5083 aluminum alloy backing plates. The 14.5 mm BS41 tungsten-carbide projectile was selected as the test rounds. Figure 6 summarizes the ballistic limit test results. One can observe that the ballistic performance of the pressureless sintered SiC almost matches (<3% difference in ballistic limit) the more expensive PAD SiC-B from Cercom at equivalent areal density and exceeds other expensive PAD SiC (>5%) and Si$_3$N$_4$ (>12%) materials from Ceradyne.

CONCLUSIONS

It is possible to manufacture relatively thick (19.1 mm) pressureless-sintered SiC that have a density of greater than 98% of theoretical. However, changes to green tile fabrication methods and the sintering profile may be necessary to eliminate low-density (and lower strength) regions in the center of thicker (27.9 mm) tiles. With proper modifications to the fabrication technique this pressureless-sintered SiC is expected to exhibit MOR values approaching 500 MPa. These changes to the manufacturing process are not expected to significantly change the projected manufacturing cost of about $140/kg for finished tiles, suitable for various armor applications.

The ballistic performance of the pressureless sintered monolithic SiC is successfully demonstrated to be equivalent to that of more expensive industrial standard ceramic armor materials. The whisker-reinforced SiC composite performs poorly due to low final density and elongated grain growth.

REFERENCES

[1]Lillo, T.M., Chu, H.S., Bailey, D.W., Harrison, W.M., Laughton, D.A, "Development of a Pressureless Sintered Silicon Carbide Monolith and Special-Shaped Silicon Carbide Whisker Reinforced Silicon Carbide Matrix Composite for Lightweight Armor Application", in the Proceedings of the 27th Annual Cocoa Beach Conference on Advanced Ceramics and Composites, January 2003, to be published.

[2]Liou, W.J., "Stress Distributions of Short Fiber Composite Materials", Computer & Structures, vol. 62 (1997), pp. 999-1012

[3]Zhu, Y.T., Valdez, J.A., Beyerlein, I.J., Zhou, S.J., Liu, C., Stout, M.G., Butt, D.P. and Lowe, T.C., "Mechanical Properties of Bone-Shaped-Short-Fiber Reinforced Composites", Acta Mater., vol. 47 (1999), pp. 1767-1781.

[4]Outokumpu HSC Chemistry for Windows, software version 1.10, Outokumpu Research Oy, Finland.

[5]German, R.M., Powder Metallurgy Science, Metal Powder Industries Federation, Princeton, NJ, 1984, Chapter 5.

[6]Zhan, G., Xie, R., Mitmo, M. and Kim, Y, "Effect of β-to-α Phase Transformation on the Microstructural Development and Mechanical Properties of Fine-Grained Silicon Carbide Ceramics", J. Am. Ceram. Soc., vol. 84 (2001), pp. 945-950.

ACKNOWLEDGEMENTS

Prepared for the U.S. Department of Energy through the INEEL LDRD Program under DOE Idaho Operations Office Contract DE-AC07-99ID13727.

Authors wish to thank Dr. William J. Bruchey of the Armor Mechanics Branch of the Army Research Lab for performing the ballistic limit tests.

DESIGN AND MANUFACTURING B$_4$C-SiC LAYERED CERAMICS FOR ARMOR APPLICATIONS

Nina Orlovskaya
Drexel University
3141 Chestnut Str.
Philadelphia, PA 19104

M. Lugovy, V. Subbotin, O. Rachenko
Institute for Problems of Materials Science
3 Krzhizhanovski Str.
Kiev, 03142, Ukraine

J. Adams
Army Research Laboratory
Aberdeen Proving Ground
MD 21005

M. Chheda, J. Shih
Ceradyne Inc.
3169 Redhill Ave.
Costa Mesa, CA 92626

J. Sankar, S. Yarmolenko
North Carolina A&T State University
1601 E. Market Str.
Greensboro, NC 27411

ABSTRACT

Boron carbide-silicon carbide ceramic composites are very promising armor materials because they are intrinsically very hard. However, their fracture toughness is not very high. Their ballistic properties could be significantly increased if the brittleness of these materials could be decreased. Here we report a development of boron carbide-silicon carbide layered ceramics with controlled compressive and tensile stresses in separate layers. The layered composites were designed in a way to achieve high compressive residual stresses in thin B$_4$C-SiC based layers and low tensile residuals stresses in thick B$_4$C layers. The residual stresses were controlled by the phase composition of layers and the layers thickness. The B$_4$C-30wt%SiC/B$_4$C laminates were manufacture based on the optimized design for high apparent fracture toughness. The laminates' processing included milling of powders, rolling, and hot pressing steps. Such B$_4$C-SiC laminates with strong interface can provide high apparent fracture toughness and damage tolerance along with high protection capabilities. The work is in the progress to measure the fracture toughness of laminates, as well as their strength, hardness and the ballistic performance.

INTRODUCTION

Ceramics offer a number of attractive properties. These include high specific stiffness, high specific strengths, low thermal conductivities, and chemical inertness in many environments. Ceramics and ceramic composites are attractive materials for use in armor systems due to low density, superior hardness, and compressive strength values relative to metals. As a result, ceramics have been subjected to a multitude of ballistic and dynamic behavior investigations [1]. However, the widespread usage of ceramics is currently hampered by their lack of the requisite toughness. The latest developments in ceramic composites show that the use of layered materials is perhaps the most promising method to control cracks and brittle fracture by deflection, microcracking, or internal stresses [2-4]. Laminates with strong interfaces, combined with excellent fracture toughness and damage tolerance, can potentially provide the highest ballistic performance. The way to achieve the highest possible fracture toughness is to control the level of residual stresses in the separate layers. It is also a way to increase the failure strength of ceramics by creating a layer with compressive stresses on the surface that will arrest the surface cracks and achieve higher failure stresses [5]. The variable layer composition, as well as the system's geometry, allows the designer to control the magnitude of the residual stresses in such a way that compressive stresses in the outer layers near the surface increase strength, flaw tolerance, fatigue strength, fracture toughness and stress corrosion cracking. In the case of symmetrical laminates, this can be done by choosing the layer compositions such that the coefficient of thermal expansion (CTE) in the odd layers is smaller than the CTE of the even ones. The changes in compressive and tensile stresses depend on the mismatch of CTE's, Young's moduli, and on the thickness ratio of layers (even/odd). However, if the compressive stresses exist only at or near the surface of ceramics and are not placed inside the material, they will not effectively hinder internal cracks and flaws [6].

Boron carbide is an important ceramic material with many useful physical and chemical properties. After cubic boron nitride, it is the hardest boron containing compound [7]. Its high melting point, high elastic modulus, large neutron capture section, low density, and chemical inertness make boron carbide a strong candidate for several high technology applications. Due to its low density and superior hardness, boron carbide is a very promising material for light-weight ballistic protection. Boron carbide exists as a stable single phase in a large homogeneity range from B_4C to $B_{10.4}C$ [8]. The most stable boron carbide structure is rhombohedral with a stoichiometry of $B_{13}C_2$, $B_{12}C_3$, and some other phases close to $B_{12}C_3$ [9]. The Vickers hardness of B_4C is in the range of 32-35 GPa [10]. There is an indication that hardness of stoichiometric B_4C is the highest one in comparison with boron rich or carbon rich boron carbide compositions [11-13]. However, B_4C-based composites have a relatively low fracture toughness of 2.9-3.5MPam$^{1/2}$ [14]. While high hardness is one of the very important requisite indicators for a material's ballistic potential, toughness might play an equally important role. Only materials with both high hardness and high fracture toughness are expected to yield the desired high ballistic performance [1, 15]. Therefore, a significant increase in fracture toughness of boron carbide based laminates has the potential for realization of significantly improved armor material systems.

Brittleness of boron carbide ceramic laminates can be controlled by designing the distribution of residual stresses, i.e. placing the layers with high compressive stresses into the bulk of the material. The sign and value of the bulk residual stresses have to be firmly established by theoretical prediction [16]. A significant increase in ballistic protection of B₄C based laminates can be achieved by designing high compressive stresses placed into the materials bulk. The goal of this research was to develop the design and manufacturing process of boron carbide-silicon carbide ceramic laminates with controlled residual stresses. In this article we demonstrate a laminate design concept by determining the prospective combination of layers, their geometry and microstructure for the B₄C/B₄C-30wt%SiC system, as well as a laminates' manufacturing route. Mechanical properties, such as Young's modulus, fracture toughness, hardness, and a ballistic performance of the developed laminates will be reported elsewhere.

THERMAL RESIDUAL STRESSES AND THEIR CALCULATION

In this work the two-component brittle layered composites with symmetric macrostructure are considered. The layers consisting of different components alternate one after another, but the external layers consist of the same component. Thus, the total number of layers N in such a composite sample is odd. The layers of the first component including two external (top) layers are designated by index 1 (j = 1), and the layers of the second component (internal) are designated by index 2 (j = 2). The number of layers designated by index 1 is $(N+1)/2$, and the number of layers designated by index 2 is $(N-1)/2$. The layer of each component has some constant thickness, and the layers of same component have identical thickness.

There are effective residual stresses in the layers of each component in the layered ceramic composite. During cooling, the difference in deformation, due to the different thermal expansion factors of the components, is accommodated by creep as long as the temperature is high enough. Below a certain temperature, which is called the "joining" temperature, the different components become bonded together and internal stresses appear. In each layer, the total strain after sintering is the sum of an elastic component and of a thermal component [17, 18]. The residual stresses in the case of a perfectly rigid bonding between the layers of a two-component material are [18]:

$$\sigma_{r1} = \frac{E_1' E_2' f_2 (\alpha_{T2} - \alpha_{T1}) \Delta T}{E_1' f_1 + E_2' f_2} \qquad (1)$$

and

$$\sigma_{r2} = \frac{E_2' E_1' f_1 (\alpha_{T1} - \alpha_{T2}) \Delta T}{E_1' f_1 + E_2' f_2}, \tag{2}$$

where $E_j' = E_j / (1 - \nu_j)$, $f_1 = \frac{(N+1)l_1}{2h}$, $f_2 = \frac{(N-1)l_2}{2h}$, E_j and ν_j are the elastic modulus and Poisson's ratio of j-th component respectively, l_1 and l_2 are the thickness of layers of the first and second component, α_{T1} and α_{T2} are the thermal expansion coefficients (CTE) of the first and second components respectively, ΔT is the difference in temperature of joining temperature and current temperature, and h is the total thickness of the specimen.

Equations (1) and (2) give the residual stresses in layers, which have an infinitive extent. Far away from the free surface, the residual stress in the layer is uniform and equibiaxial. In the bulk of layers, the stress perpendicular to the layers is zero. At the free surface of the laminates, the stresses are different from the bulk stresses. Near the edges, the residual stress state is not equibiaxial because the edges themselves must be traction-free. Highly localized stress components perpendicular to the layer plane exist near the free surface and it decreases rapidly from the surface becoming negligible at a distance approximately on the order of the layer thickness. These stresses have a sign opposite to that of the equibiaxial stresses deep within the layer. Therefore, if the bulk stress is compressive within the material, the tensile stress components appear at or near the free surface of a layer.

LAMINATE DESIGN FOR ENHANCED FRACTURE TOUGHNESS

The schematic presentation of symmetric three-layered and nine-layered composites that were considered for a design and a manufacturing is shown in Fig.1. The proposed design targeted a fracture toughness increase of B_4C-SiC composites and was based on the preliminary results both from our work [19, 20] and from the work of others [9, 21-24].

In case of non-homogeneous (particularly, layered) materials, so-called apparent fracture toughness should be considered. This is the fracture toughness of some effective homogeneous specimen. If we measure fracture toughness in bending, the effective sample parameters should satisfy the following conditions: 1) the specimen has to have the same dimensions as a real layered specimen; 2) the notched sample has a notch depth equal to that of the real layered specimen; 3) under the same loading conditions the specimen has to demonstrate the same load to fracture as that of the real layered specimen. Under these considerations the apparent fracture toughness is the fracture toughness calculated from a testing data of the layered sample considering this specimen as "homogeneous". Such an approach does not meet the fracture mechanics requirement of taking into account all features of stress distribution near crack tip in layered media, but it is still a useful characteristic allowing an effective contribution of such factors as residual stresses and a material inhomogeneity to be accounted for.

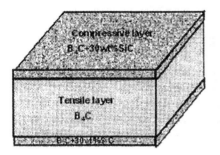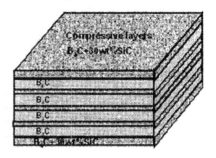

Fig. 1. Schematic presentation of symmetric 3 layered and 9 layered composite.

The compressive residual stress σ_r in the top layers of a laminate shields natural and artificial cracks in the layer. Therefore, the effective (apparent) fracture toughness of such a structure increases. The more compressive residual stress induces, the more shielding occurs. Another important factor that contributes to the apparent fracture toughness increase is a crack length a. A longer crack promotes more shielding. A maximum length of a transverse crack in a top compressive layer is limited by the layer thickness l_1. These two factors determine the apparent fracture toughness of the material.

In general, a condition of a crack growth onset is $K_a+K_r=K_c$, where $K_a=K_a(\sigma_a, a)$ is the applied stress intensity factor that can be measured, σ_a is the distribution of applied stress resulted from bending, $K_r=K_r(\sigma_r, a)$ is the stress intensity factor due to a residual stress, and K_c is the intrinsic fracture toughness of a material in the layer. If a condition of a crack growth onset is fulfilled then $K_a=K_c-K_r$ is the apparent fracture toughness. If σ_r is compressive, then $K_r<0$ and K_a increases. The more $|\sigma_r|$, the more K_a. The more a, the more K_a. The largest value of a crack length in compressed layer is l_1. The maximum apparent fracture toughness can be obtained for such crack. Unfortunately, small cracks have K_a close to K_c.

The design technique to obtain the enhanced fracture toughness of a layered composite is as follows. First, the compositions of layers are selected depending on a future application of the composite. Then, the relevant material constants entering the design are determined. The constants for design are the coefficient of thermal expansion, Young's modulus, Poisson's ratio, and a density of the corresponding constituents. A very important but experimentally unknown parameter is ΔT – a "joining" temperature. Further, effective coefficients of thermal expansion, an effective Young's modulus, an average density and a thickness ratio of layers are determined using the rule of mixture. The next step of design is the selection of the layer's number. It can be any appropriate number depending on the required total thickness of the tile. To obtain the enhanced fracture resistance of layered composite, the factors affecting the apparent fracture toughness should be taken into account. Usually, the thickness of the thinnest possible layer is limited by the manufacturing technology. Note that a compressive layer should be thin enough to reach high level of residual stress. Another important requirement is a thickness ratio of layers with high CTE (a tensile stress) and layers with lower CTE (a compressive stress). Any appropriate thickness ratio can be used as a first approximation. Then tensile layer thickness is found. After this, the calculation of residual stresses is fulfilled using (1) and (2). The total thickness of the sample is also determined at this step for a given layer's thickness ratio taking into account the selected number of layers. The thickness ratio is changed after analysis of the residual stress and the total thickness of the specimen. Note that increasing ratio of tensile layer thickness to compressive layer thickness decreases tensile residual stress. However, it can result in increasing total thickness of sample. After changing thickness ratio, the calculation is repeated. Such iterations are continued to find a unique optimal layer thickness ratio that produces the maximum possible compressive residual stress, low tensile residual stress, and necessary total thickness of the sample. The maximum possible apparent fracture toughness of the corresponding layered structure is also determined in all iterations as an indicative parameter of the design. The determination of the apparent K_{Ic} uses the compressive residual stress and the thickness of a top layer as a crack length at any given iteration. These two parameters (the compressive residual stress and the thickness of the top layer) have trends acting in opposite directions. A decrease in the top layer thickness can increase the residual stress in the layer, but it decreases the length of the maximum crack. Therefore, the maximum apparent fracture toughness was always used to analyze the correct thickness ratio.

MANUFACTURING OF LAMINATES

The material systems selected for the proposed study were B_4C and B_4C-30wt%SiC because of their promise for ballistic applications [25, 26]. Table 1 shows the relevant material constants entering the design (compiled from literature), and Tables 2 and 3 show the corresponding calculated residual stresses in the B_4C/B_4C-30wt%SiC laminates. The maximum possible apparent fracture toughness for corresponding layered structures is also presented in the Tables 2 and 3. The layers under tensile stress have higher CTE, and in this case they are B_4C layers. The layers under compressive stress have lower CTE; here they are B_4C-30wt%SiC layers. A temperature T= 2150°C was used for the majority of the calculations, when residual stresses appeared in the layers upon cooling from the hot pressing temperature. There is no liquid phase present during the sintering of B_4C/B_4C-SiC ceramics [27], therefore, the hot pressing temperature was used as a "joining" temperature ΔT for calculations. It should be noted that all laminates were designed in such a way that the tensile stresses had been maintained at low values.

Table I. Properties of ceramics used in the stress calculation

Composition	E, Gpa	Poisson's ratio	CTE, 10^{-6} K^{-1}
B_4C	483	0.17	5.5
SiC	411	0.16	3

Table II. Three layered composite design. A total thickness of a tile – 10.5mm

Composition	Thickness of Layers, μm		σ_{comp},	σ_{tens},	Apparent K_{Ic}, $MPam^{1/2}$
	B_4C-30wt%SiC	B_4C	MPa	MPa	
B_4C-30wt%SiC/B_4C	900	8700	632	131	44

Table III. Nine layered composite design. A total thickness of tile – 10.35mm.

Composition	Thickness of Layers, μm		σ_{comp},	σ_{tens},	Apparent K_{Ic}, $MPam^{1/2}$
	B_4C-30wt%SiC	B_4C	MPa	MPa	
B_4C-30wt%SiC/B_4C	150	2250	662	99	32

B_4C and α-SiC powders with a grain size of 2-5 μm were used for laminates manufacturing. For B_4C-30wt%SiC batch preparation, the wet mixture in acetone was done in the polyethylene bottle and B_4C balls as grinding media for 48 hours. The laminates were produced via rolling followed by hot pressing techniques. The formation of a thin ceramic layer is of specific importance, as the sizes of residual stress zones (tensile and compressive) are directly connected with the thickness of layers. The advantage of rolling, as a method of green layers production, includes easy thickness control, high green density of the tapes, and a rather low amount of solvent and organic additives are necessary in comparison with other methods like tape casting. Additional powder refinement, giving a higher sintering reactivity, might occur due to large forces applied in the pressing zone during rolling. In our case there is a challenging problem to produce thin tapes with a small amount of plasticizer and sufficient strength and elasticity to handle green layers after rolling. Crude rubber (1-3wt%) has to be added to the mixture of powders as a plastisizer through a 3 % solution in petrol. Then the powders were dried up to the 2wt% residual amount of petrol in the mixture. After sieving powders with a 500 μm sieve, granulated powders were dried up to the 0.5wt% residual petrol. The schematic presentation of rolling is shown in Fig. 2.

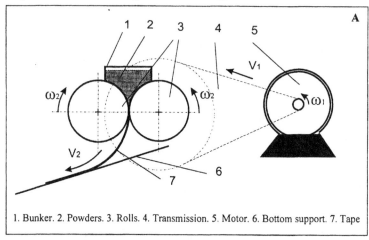

1. Bunker. 2. Powders. 3. Rolls. 4. Transmission. 5. Motor. 6. Bottom support. 7. Tape

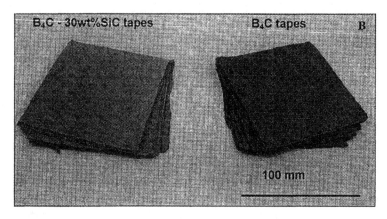

Fig. 2. A) Schematic presentation of rolling. B) Photograph of B_4C and B_4C-30wt%SiC rolled tapes. The thickness of an individual tape after rolling is between 0.4-0.5mm.

The powders are continuously supplied in the bunker and further into the deformation zone in between rolls. The powders are supplied to the deformation zone due to both the gravitational force and the friction force between rolls and powders. The relative density of the tape (ρ_r) can be calculated from

$$\rho_r = \frac{\rho_p}{\lambda}(1 + \frac{\alpha^2 R}{h_s}) \qquad (3)$$

where ρ_p is a relative powder density, λ is a drawing coefficient, α is an intake angle, and R is a roll diameter. A roll mill with 40mm rolls was used for rolling. The velocity of rolling was in the range of 1-1.2 m/min. Working pressure was varied from 0.1 t/cm^2 for relative density of tapes 64% to 1 t/cm^2 for 74% density.

Samples of ceramics were prepared by hot pressing of the rolled tapes stacked together. The hot pressing conditions were as follows: a) a heating rate was 100°C/min; b) a hot pressing temperature was kept at 2150°C during hot pressing of a majority of the tiles, and some hot pressing was done at 2200°C to ensure that fully dense materials were obtained; c) a pressure was kept at the level of 30 MPa; and d) a dwelling time at hot pressing temperature was 50-60 minutes. Graphite dies were used for the hot pressing of laminates with graphite surfaces coated by BN layer in order to prevent a direct contact between graphite and ceramic material. 90x90x10mm tiles were produced as a result of hot pressing. Dense (97-100% of density) laminate samples were obtained.

MICROSTRUCTURE OF LAMINATES

During hot pressing of laminates the shrinkage (in 3 times) of the individual layers occurred, and their thickness become 0.15 mm after hot pressing. The interfaces between individual layers of the same composition completely disappeared and only the interface between B$_4$C-30wt%SiC and B$_4$C layers could be distinguished.

A fracture surface of a three-layer tile hot pressed at 2200°C for 1 hour is shown in Fig.3. The layered composite demonstrates typical brittle fracture. One can see the interface between the B$_4$C-30wt%SiC outer layer and the pure B$_4$C inner layer (Fig.3a). The fracture surface of the B$_4$C layer is presented in Fig.3b. Fig.3c shows the fracture surface of the B$_4$C-30wt%SiC layer. The cleavage steps on the B$_4$C fracture surface are presented in Fig.3d. Such cleavage mode plays an important role both in fracture and in the fragmentation event during ballistic impact [28].

As one can see from Fig.3, the B$_4$C grain size in B$_4$C-30wt%SiC layers was in the range of 4-6 μm, SiC grain size was in the range of 2-5 μm. B$_4$C grain size in pure B$_4$C layers could not be calculated because of a pure transgranular fracture mode with no grains or grain boundaries revealed after fracture. The significant grain growth of boron carbide is expected during hot pressing at 2200°C. However, in B$_4$C-30wt%SiC layers, the existence of the SiC phase prevented the exaggerated grain growth and the grain size distribution was homogeneous. Tiles hot pressed at 2200°C for 1 hour were fully dense. Tiles hot pressed at 2150°C for 30 or 45 minutes contained some amount of porosity (2-5%) that was concentrated along the interfaces and mostly in pure B$_4$C layers. Such porosity could be detrimental for material hardness, affecting Young's modulus and density, thus significantly lowering the ballistic properties of the laminates.

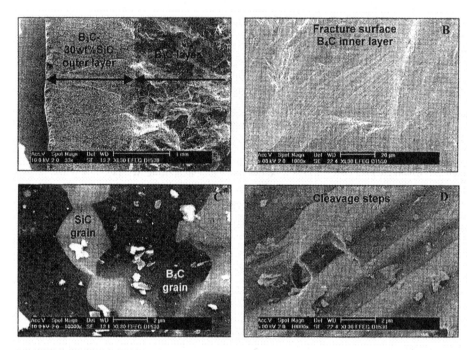

Fig. 3. Fracture surface of a three layered tile. A) An interface between B$_4$C-30wt%SiC outer layer and pure B$_4$C inner layer; B) A fracture surface of B$_4$C layer; C) A fracture surface of B$_4$C-30wt%SiC; D) Cleavage steps on the B$_4$C fracture surface

CONCLUSIONS

Boron carbide-silicon carbide ceramics have been used in the design and manufacturing of three layered and multilayered composite with strong interfaces for enhanced fracture toughness. The model of heterogeneous layered system was used to develop optimal design parameters. As a result, the laminates with high compressive residual stresses (up to 650MPa) and low tensile residual stresses (below 150 MPa) were developed. The feasibility of designing laminate composite systems with enhanced toughness by incorporation of thin layers with high compressive stresses in the ceramics was demonstrated. Both three-layered and nine-layered B$_4$C-30wt%SiC/B$_4$C composites were manufactured using rolling and hot pressing techniques. The work is currently in progress to study the

Ceramic Armor and Armor Systems

mechanical properties, such as fracture toughness, strength, hardness, as well as ballistic performance of developed B_4C-30wt%SiC/B_4C ceramic laminates.

This research represents a first step in laminate ceramics development that provides superior ballistic protection. The method of designing the residual stresses developed in this work can be also applied for the broad range of layered material systems. The results of this study are likely to find practical applications in the field of ballistic protection, mechanical behavior, lifetime prediction, and reliability of advanced ceramic composites.

ACKNOWLEDGEMENT
This work was supported by AFOSR, the project # F49620-02-0340 and by the European Commission, the project ICA2-CT-2000-10020, Copernicus – 2 Program. This work was also partly performed at the Army Center for Nanoscience and Nanomaterials, North Carolina A&T State University.

REFERENCES
[1] J.S.Sternberg, "Materials properties determining the resistance of ceramics to high velocity penetration," Journal of Applied Physics, 69 3417-3424, (1989).

[2] W.J.Clegg, "Controlling cracks in ceramics", *Science,* **286** 1097-1099 (1999).

[3] M.Lugovy, N.Orlovskaya, K.Berroth, J.Kuebler, "Microstructural Engineering of Ceramic-Matrix Layered Composites: Effect of Grain Size Dispersion on Single-Phase Ceramic Strength," *Composite Science and Technology* **59** [2] 283-289 (1999).

[4] M.Lugovy, N.Orlovskaya, K.Berroth, J.Kuebler, "Macrostructural engineering of ceramic-matrix layered composites," *Composite Science and Technology,* **59** [8] 1429-1437 (1999).

[5] P.Honeyman-Colvin, F.F.Lange, "Infiltration of porous alumina bodies with solution precursors: strengthening via compositional grading, grain size control, and transformation toughnening," *Journal of the American Ceramic Society,* **79** 1810-1814 (1996).

[6] M.P.Rao, A.J.Sanchez-Herencia, G.E.Beltz, R.M.McMeeking, F.F.Lange, "Laminar ceramics that exhibit a threshold strength," *Science,* 286 102-104 (1999).

[7] I.J.McColm, Ceramic Hardness, Plenum Press, New York, 1990.

[8] K.A.Schwetz, A.Lipp, Ulmann's Encyclopedia of Industrial Chemistry, A4, VCH, 295, 1981.

[9] E.Amberger, W.Stumpf, K.-C.Buschbeck, Handbook of Inorganic Chemistry, 8th ed., Springer-Verlag, Berlin 1981.

[10] F.Thevenot, "Boron carbide – A comprehensive review," *Journal of the European Ceramic Society,* **6** 205-225 (1990).

[11] B.Champagne, R.Angers, "Mechanical properties of hot pressed B-B_4C materials," *Journal of the American Ceramic Society,* **62** 149-153 (1979).

[12]K.Niihara, A.Nakahira, T.Hirai, "The effect of stoichiometry on mechanical properties of boron carbide," *Journal of the American Ceramic Society,* **67** C13-C14 (1984).

[13]F.Trevenot, in G. de With, R.A.Terpstra, R.Metselaar (Eds.), Properties of Ceramics, Elsevier Appl. Sci., London and New York, 1989.

[14]H.Lee, R.Speyer, "Hardness and fracture toughness of presureless sintered boron carbide," *Journal of the American Ceramic Society,* **85** 1291-1293 (2002).

[15]M.L.Wilkens, "Mechanics of penetration and perforation," *International Journal of Engineering Science,* **16** 793-807 (1978).

[16]C.A.Folsom, F.W.Zok, F.Lange, "lexural properties of brittle multiplayer materials: I. Modeling," *Journal of the American Ceramic Society,* **77** 689-696 (1994).

[17]S.Ho, C.Hillman, F.F.Lange, Z.Suo, "Surface cracking in layers under biaxial compressive stress," *Journal of the American Ceramic Society,* **78** 2353-2359 (1995).

[18]T.Chartier, D.Merle, J.L.Besson, "Laminar ceramic composites," *Journal of the European Ceramic Society,* **16** 101-107 (1995).

[19]V.Yaroshenko, N.Orlovskaya, M.-A.Einarsrud, V.Kovylayev, "Processing of multilayered Si_3N_4-TiN hot-pressed ceramic composites,"; pp. 285-295 in *Multilayered and fibre-reinforced composites: Problems and Prospects,* Edited by Hui Haddad. Kluwer, Dordrech, 1998.

[20]M.Lugovy, N.Orlovskaya, V.Slyunayev, G.Gogotsi, J.Kuebler, A.Sanchez-Herencia, "Crack bifurcation features in laminar specimens with fixed total thickness," *Composite Science and Technology,* **62**, 819-830, (2002).

[21]M.Oechsner, C.Hillman, F.Lange, "Crack bifurcation in laminar ceramic composites," *Journal of the American Ceramic Society,* **79** 1834-1838 (1996).

[22]D.Green, P.Cai, G.Messing, "Residual stresses in alumina-zirconia laminates," *Journal of the European Ceramic Society,* **19** 2511-2517 (1999).

[23]R.Lakshminarayanan, D.K.Shetty, R.A.Cutler, "Toughening of layered ceramic composites with residual surface compression," Journal of the American Ceramic Society, **79** 79-87 (1996).

[24]A.Blattner, R.Lakshminarayanan, D.K.Shetty, "Toughening of layered ceramic composites with residual surface compression: effect of layer thickness," *Engineering Fracture Mechanics,* **68** 1-7 (2001).

[25]C.J.Shih, M.A.Meyers, V.F.Nesterenko, S.J.Chen, "Damage evolution in dynamic deformation of silicon carbide," *Acta Materialia,* **48** 2399-2420 (2000).

[26]D.L.Orphal, R.R.Franzen, A.C.Charters, T.L.Menna, A.J.Piekutowski, "Penetration of confined boron carbide targets by tangsten long rods at impact velocities from 1.5 to 5.0 km/s," *International Journal of Impact Engineering,* **19** [1] 15-29 (1997).

[27]P.S. Kislyi, M.A. Kuzenkova, N.I. Bondaruk, B.L. Grabchuk, Boron Carbide, Naukova Dumka, Kiev, 1988, In Russian.

[28]M. Chen, J. W. McCauley, K.J. Hemker, "Shock induced localized amorphization in boron carbide," *Science,* **299** 1563-1565 (2003).

IMPROVED PRESSURELESS DENSIFICATION OF B$_4$C

Robert F. Speyer
School of Materials Science
  and Engineering
Georgia Institute of Technology
771 Ferst Drive
Atlanta, GA 30332-0245

Hyukjae Lee
Department of Aeronautics
  and Astronautics
Air Force Institute of Technology
Wright-Patterson AFB, OH 45433-
7765

ABSTRACT

B$_4$C powder compacts were sintered using a graphite dilatometer in flowing He under constant heating rates. With 3 wt% carbon doping, the onset of densification was lowered from 1800 to 1350°C, attributed to reduction-removal of B$_2$O$_3$ coatings which had inhibited contact between B$_4$C particles. Limited particle coarsening, attributed to the presence or evolution of the oxide coatings on un-doped specimens, occurred in the range 1870-1950°C. In the temperature range 2010-2140°C, densification was concurrent with evaporation-condensation coarsening. Above 2140°C, rapid densification ensued, which was interpreted to be the result of the formation of a eutectic grain boundary liquid, or activated sintering facilitated by non-stoichiometric volatilization of B$_4$C, leaving carbon behind. Rapid heating through temperature ranges in which coarsening occurred fostered increased densities. Soaking in H$_2$ at 1350°C extracted B$_2$O$_3$ coatings from B$_4$C particles, permitting a lower temperature onset of sintering for undoped specimens, and attenuating a mechanism of coarsening. Remnant H$_2$ needed to be removed from the furnace while specimens were heated through temperature ranges in which evaporation-condensation coarsening competed with sintering (2010-2140°C), since its presence increased the B$_4$C vapor pressure. Heat treatment of B$_4$C compacts in a 50-50 H$_2$-He mixture at 1350°C, followed by a purge of the H$_2$ gas, and then rapid heating to 2230°C resulted in a percentage of theoretical density of 94.7.

INTRODUCTION

Boron carbide (B$_4$C) is a covalently-bonded solid with a high melting point (2427°C), extremely high hardness (Vickers: 3770 kg/mm^2), a low density (2.52 g/cm^3) and a high neutron absorption cross-section [1]. It is a solid solution with carbon in the range 8.8-20.0 mol%. Boron carbide has been used for light weight ceramic armor, for wear-resistant components such as blasting nozzles and grinding wheels, and for control rods in nuclear reactors. Its synthesis and properties have been reviewed elsewhere [1-7].

Sintering of pure boron carbide to high densities has proven difficult. Specific additives (sintering aids such as carbon, Al_2O_3 and TiB_2) or hot pressing have been used to achieve near full density. B_4C powders are typically hot-pressed at ~2100°C and 30-40 MPa to obtain dense articles [8-9]. Post hot isostatic pressing (HIP) has been applied to carbon-doped B_4C in order to reach 100% of theoretical density, which has led to improvements in the flexural strength, modulus of elasticity, and wear resistance of the final product [10].

Hot pressing is only applicable to rather simple shapes. Pressureless sintering of B_4C is often preferable in order to avoid the expensive diamond machining required to form complex shapes. The best known additive for pressureless sintering of B_4C is carbon. Carbon additions have usually been through in-situ pyrolysis of a Novolaque-type phenolic resin which serves as a pressing agent and later yields ~40 wt% carbon upon decomposition.

The mechanisms responsible for the limited densification of B_4C has not been firmly established. Dole et al. observed highly coarsened sintered compacts of undoped B_4C after heat-treating at 2000°C, whereas carbon-doped samples had not yet undergone much coarsening [11-13]. The authors suggested that B_2O_3 liquid in the starting B_4C powder may have provided a rapid diffusion path along particle surfaces, facilitating particle coarsening. They also suggested that the carbon additions could eliminate B_2O_3 from boron carbide compacts [11,12]. In their work, particle coarsening started as low as 1750°C, and after further heat-treatment, a dense polycrystalline solid, surrounded by large interconnected pores formed. The pores became stable and acted as densification barriers; large grains and large pores increase the diffusion distances to fill pores.

It is of fundamental and practical interest to establish why non oxide covalently-bonded compounds such as B_4C do not sinter to high density without additives under pressureless sintering. Since residual carbon reduces the strength of B_4C, it is of interest to facilitate optimum densification without carbon additions. The focus of this study was to investigate the competing processes which attenuate sintering driving forces, and once established, investigate methods to minimize them.

EXPERIMENTAL PROCEDURE

Commercially available B_4C powders (Grade HS, H. C. Starck, Berlin, Germany) were used in the as-received state. Table I shows the characteristics of the powder based on the manufacturer's data. For the 3 wt% carbon additions, phenolic resin solution (SP6877, Schenectady International, Schenectady, NY) was dissolved in acetone, to which the B_4C powder was added. The mixed solution was dried while stirring. After further drying, the products were ground in a mortar and screend to -125 mesh. Powders were uniaxially pressed (~5 mm in height and 6.4 mm in diameter) into compacts in a steel die at 200 MPa. Green densities, calculated based on specimen dimensions and mass, were in the range of 67-68% of theoretical density for undoped powders, and 70-71% of theoretical density for carbon-doped powders.

Weight losses of B_4C powders upon heating were measured in part via thermogravimetric analysis (TGS-2 analyzer and balance control unit, Perkin-Elmer, Norwalk, CT). The TG chamber was twice evacuated and backfilled with argon (ultra-high purity). Raw material powder, placed on Pt specimen pans, was suspended by a thin platinum wire into the hot zone of a small Pt-wound furnace. A heating rate of 4°C/min to 380°C was used. In a separate series of experiments, weight losses up to 1910°C were evaluated by heating pressed pellets, in flowing He, in the dilatometer (discussed in the following paragraphs) to specified temperatures, followed by furnace cooling (heating elements turned off), still under flowing He. Specimens were weighed immediately after removal from the furnace, and also weighed after exposure to room air for one day.

The furnace (Model 1000-2560-FP, Thermal Technology Inc., Santa Rosa, CA) used for heat-treatments was comprised of graphite heating elements and fibrous insulation. Inert and reducing environments are appropriate for the furnace. At 1750°C, a 5 cm long temperature zone with a uniformity tolerance of ±3°C has been specified by the manufacturer. Temperature was monitored using a pyrometer (Model MA1SC, Raytek Co., Santa Cruz, CA) measuring at a wavelength of 1.0 μm through a fused silica viewing port mounted on the furnace radial wall.

A double-pushrod dilatometer (Theta Industries Inc., Port Washington, NY) extended into the furnace, which used a linear variable differential transformer (LVDT) position transducer to measure specimen expansion/shrinkage. Both the cylindrical casing and the sample pushrod were made of graphite (Poco Graphite Inc., Decatur, Texas). A counter-weight was applied to the sample pushrod to avoid particle sliding within the green compact, imposed by the force of the pushrod. A He atmosphere was created within the furnace chamber through repeated (3×) evacuation via a mechanical pump, followed by backfilling from compressed gas cylinders. He, He/H_2, or H_2 atmospheres were used. The gas flow rates during heating were ~0.5 liters/min (furnace vessel interior volume: 3 liters). He/H_2 flow and mixture were controlled by independent mass flow controllers connected to tanks of each pure compressed gas.

Table I B_4C Powder characteristics.

Surface area: 18.8 m²/g Particle size: 90% of particles ≤ 2.99 μm 50% of particles ≤ 0.84 μm 10% of particles ≤ 0.24 μm Total boron: 75.39 wt% Total carbon: 22.26 wt% B/C molar ratio: 3.76	Impurity levels: 1.500 wt% O 0.410 wt% N 0.022 wt% Fe 0.055 wt% Si 0.003 wt% Al 0.230 wt% Other

During densification, the specimen was not only shrinking due to sintering but was also expanding due to thermal expansion from rising specimen temperature. This introduced inconsistencies in shrinkages measured at room temperature after firing. As a correction, true specimen contractions were continuously recalculated during heat-treatment by subtracting the expansion profile of a fully sintered sample from the measured displacement. Another source of inconsistent terminal density was shrinkage due to continued sintering which occurred during the early part of the cooling cycle. To minimize this, the setpoint was dropped instantaneously by 100°C, followed by cooling at a constant rate of 35°C/min [14].

The maximum possible densifications of undoped and carbon-doped samples as a function of heating rate were studied. Samples were heated at rates of 10, 30 and 100°C/min up to 2250°C and then soaked until the shrinkage rate reached 0.00%/min. In order to investigate the progression of weight change and particle/grain size during heat treatment, eight undoped samples were heated at 5°C/min from 1800°C to 1800, 1870, 1910, 1950, 1980, 2010, 2070, or 2130°C, and then cooled (in the furnace with no power to the heating elements) to room temperature.

After sintering, bulk density was measured using Archimedes' method. The theoretical density of the carbon-doped sample was measured using pycnometry (03-240A, Fisher Scientific, Pittsburgh, PA). The evaluated powder had been fired at 2250°C, and crushed after firing.

Sintered samples were mounted in an acrylic compound (Ultra-Mount, Buehler, Lake Bluff, IL) and ground using a rotating grinder (Ecomet-3, Buehler) and electrolytically etched. Micrographs were scanned for the purpose of grain size measurement. Software was coded to determine grain size using the linear intercept method. The mean grain size was calculated from the mean of 1000 measured intercept lengths between grain boundaries, multiplied by 1.56 [15]. For fracture surface observations (used when specimens were not adequately dense for polished-section evaluation), samples were notched and sheared to fracture. These fractured specimens were then sputter-coated with gold and mounted on aluminum stubs using silver paste to make them electrically conductive in the SEM (500S, Hitachi, San Diego, CA).

A PW 1800 powder X-ray diffractometer (Philips, Mahwah, NJ) was used to identify the phases of the starting powder and sintered compacts. The operating voltage and current were 40 kV and 30 mA, respectively. The X-ray data were collected from 20° to 90° with a scan speed of 3 s per step and a step size of 0.01°.

RESULTS AND DISCUSSION
Oxide Coatings

As shown in Figure 1, B_4C powder lost weight while being heated in argon to 380°C at 4°C/min, and did not regain this weight when cooled at the same rate, nor when held at room temperature while exposed to flowing dry air. As shown in Figure 2, specimens heated to temperatures up to 1200°C initially lost weight, but then regained more than this weight over time after exposure to room air. After heating between 1200 and 1600°C, weight loss was substantial and permanent, with the rate of weight loss slowing above 1600°C. There was no regain of weight from exposure to room air for samples heat-treated to or above ~1910°C. Weight loss began again at temperatures exceeding 2010°C.

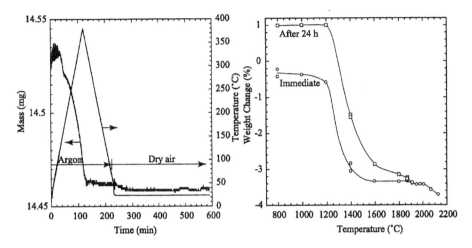

Figure 1 (left) Thermogravimetry trace of B$_4$C powder in flowing Ar under heating and cooling rates of 4°C/min.

Figure 2 (right) Weight changes in B$_4$C pressed pellets after heating to various temperatures at 50°C/min up to 1800°C, and 5°C/min above this temperature, under flowing He. Specimens were cooled in the furnace with the elements turned off under flowing He. Measurements were made immediately after removal from the furnace, and again after exposure to room air for one day.

The above phenomena are interpreted based on the presence of B$_2$O$_3$ coatings on B$_4$C particles. Based on mass loss after powder compacts were heated to 380°C, and recovery of this weight in moist air, but not dry air, the green specimens clearly absorbed atmospheric moisture. Boron oxide is highly hygroscopic at room temperature, likely forming coatings of orthoboric acid H$_3$BO$_3$. Orthoboric acid dehydrates with heating (converting to HBO$_2$ above 170°C, and H$_4$B$_4$O$_7$ at ~236°C, followed by conversion to B$_2$O$_3$ [16]). After heating to temperatures below 1200°C, the boron oxide coatings re-hydrated to an even greater extent (than their original pre-heat treatment state) when exposed to room temperature air (Figure 2). In heat treating above this temperature; however, the vapor pressure of B$_2$O$_3$ became significant. Correspondingly, a diminished weight recovery occurred during exposure to room air, since while the remaining B$_2$O$_3$ showed an affinity for atmospheric moisture, but the newly exposed B$_4$C did not. There is a lack of agreement regarding the boiling point of B$_2$O$_3$ (i.e. temperature at which its vapor pressure is 1 atm); some references indicate 1860°C [17], while others indicate 2065°C [18]. Based on the flattening of the weight loss curve in Figure 2 at 1600°C, extraction of B_2O_3 might be interpreted to be complete at this temperature. However, weight recovery in room temperature air was measured after heating up to 1910°C, implying the resilience of limited amounts of this phase approaching this temperature. Based on an average spherical particle diameter of 1 μm and a B_2O_3 weight loss of 2.4%, the average B_2O_3 coating thickness was 4.0 nm.

Particle Coarsening

Graphite was detected in the XRD pattern of the undoped compact interior sintered at 30°C/min to 2250°C; these peaks were not detected in the green compact. Comparatively more intense graphite peaks, including several lower intensity peaks which were not detected from the undoped compact, were observed from the 3 wt% carbon-doped compact after sintering at 30°C/min to 2250°C.

Early studies showed variations in final density with variations in the green density of the pressed compacts. Thus, samples used throughout the study were those which very closely matched in green density. The theoretical density of carbon-doped B_4C was taken to be 2.50 g/cm³, based on pycnometry measurements. The literature value of theoretical density was used for undoped B_4C (2.52 g/cm³).

Figures 3 and 4 show the contraction behaviors resulting from firing specimens under different heating rates. The onset of sintering of undoped specimens was at 1800°C, while for carbon-doped specimens, the onset was ~1350°C. After reaching the soaking temperature, these specimens under slower heating rates required more time to reach terminal shrinkage (set at 9.1% of corrected contraction); the sample fired at 5°C/min needed ~40 min soaking time, but only ~5 min soaking time was required for the sample exposed to a 50°C/min heating rate. Carbon-doped samples showed the opposite trend. Average grain sizes were larger for slower heating rates in both undoped and carbon-doped samples.

The maximum achieved densities of undoped and carbon-doped samples were investigated (Table II). Undoped samples were heated at different rates to 2250°C and soaked until shrinkage rates reached 0.00%/min. For these undoped samples, faster heating rates resulted in higher maximum densities and slightly smaller grain sizes. However, almost the same maximum densities and grain sizes, regardless of heating rate, were obtained for carbon-doped samples.

Figure 5 shows the progression in densification, particle/grain size and weight loss for undoped specimens heated to specified temperatures, followed by furnace cooling. Average particle/grain sizes in the figure were obtained from fracture surfaces; these values are useful only for relative comparison.

As interpreted from the figure, the vaporization of B_2O_3 coatings on the B_4C particles permitted a surge in densification between 1870 and 2010°C. During the early portion of this temperature range, i.e. 1870-1950°C, particle coarsening may be attributable to solution and precipitation of B_4C through remaining B_2O_3 liquid (coarsening removes material from small particles, and deposits it on large ones, but does not induce particle centers to approach). Alternatively, coarsening could have taken place through evaporation and condensation of rapidly evolving oxide gases (e.g. BO and CO [11]). In either case, the presence of oxide is considered to be the conduit for coarsening.

Table II: Maximum achieved densities using different heating rates to 2250°C, and soaking at that temperature until contraction rates were 0.00%. Density and grain size averages and standard deviations were based on two samples for density and three samples for grain size.

Heating rate (°C/min)	Undoped samples		Doped samples	
	Density (% of TD)	Grain size (μm)	Density (% of TD)	Grain size (μm)
10	91.33 ± 0.42	2.86 ± 0.03	98.56 ± 0.50	2.35 ± 0.02
30	92.31 ± 0.52	2.81 ± 0.04	98.65 ± 0.48	22.34 ± 0.06
100	92.76 ± 0.27	2.77 ± 0.04	98.47 ± 0.30	2.32 ± 0.03

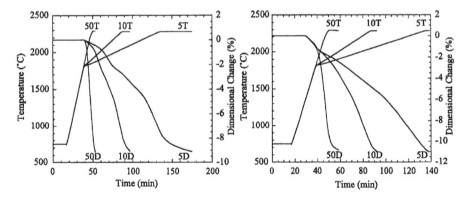

Figure 3 (left) Temperature (T) and (CTE-corrected) dimension change (D) traces of undoped samples under varying heating rates. All samples were quickly cooled when the corrected contraction reached 9.1% (90% of TD). The numbers on the figure indicate heating rates (°C/min).

Figure 4: (right) Analogous traces as in Figure 3 for carbon-doped samples. All samples were densified up to 11% (97% of TD) of corrected contraction.

Weight loss and particle/grain coarsening, stalled between 1960 and 2010°C, and resumed thereafter. Coarsening and weight loss was then concurrent with slowed densification up to ~2140°C. This is interpreted to correspond to evaporation and condensation of B₄C (the vapor pressure of B₄C is at 2138°C is reported to be 4 × 10⁻⁵ atm [19]), a coarsening mechanism typical of such cova-lently-bonded solids [20]. It is not clear whether the evaporated gaseous species were molecular B₄C, or fragments of this molecule.

Above 2140°C, densification accelerated significantly. It is difficult to at-tribute such a sharp increase in densification rate to a sudden predominance of solid state sintering over coarsening processes. Rather, grain boundary liquid may have formed at and above this temperature. This composition of B₄C solid solution fuses at ~2380°C [21]. However, the B₄C used herein contains impuri-ties such as nitrogen (in the form of BN) which may have formed a lower tem-perature eutectic liquid. As an alternative, non-stoichiometric volatilization of B₄C, leaving C behind (as indicated by XRD results), may have accelerated sin-tering via enhanced grain boundary diffusivity of boron and carbon (activated sin-tering [20]), and inhibited grain growth, keeping diffusion distances relatively short.

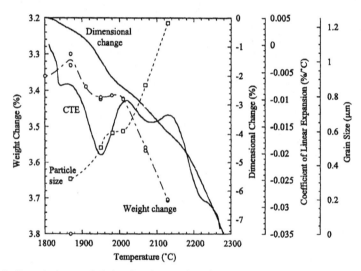

Figure 5: Trends in particle/grain size and weight change of undoped samples af-ter heating at 5°C/min up to specific maximum temperatures and then cooling with no soaking.

For doped samples, carbon, which was well distributed between particles, reacted with B_2O_3 coatings: $6C + 2B_2O_3 = B_4C + 6CO$. This allowed direct contact between B₄C particles, permitting sintering to initiate at significantly lower temperature (~1350°C, Figure 4). Thus, the significantly higher densities of carbon-doped samples in Table II are attributed, in part, to low temperature sintering via early removal of B_2O_3. Carbon at the grain boundaries may have enhanced diffusion, facilitating accelerated solid state sintering at lower tempera-tures. B₄C-C eutectic liquid forms at 2245°C [21]. Thus, eutectic liquid from

Ceramic Armor and Armor Systems

C eutectic liquid forms at 2245°C [21]. Thus, eutectic liquid from these two components, plus impurity constituents, is an alternate cause for accelerated densification starting at ~2070°C. Free carbon would also have facilitated a drag on the grain boundaries, attenuating grain growth during the later stages of sintering. Some of the improvement in final density of carbon-doped specimens must be attributed to the higher green densities of carbon-doped compacts, owing to the lubricating effects of phenolic resin.

Under rapid heating through the range 1870-1950°C (though this range shifts somewhat with heating rate), the extraction rate of B_2O_3 was accelerated via rapid heating, leaving less time for oxide-facilitated particle coarsening to take place. Similarly, rapid heating through the range 2010-2140°C minimized the time over which coarsening could occur by evaporation and condensation of B_4C. Rapid heating brought comparatively small, high surface energy particles into an elevated temperature range at which liquid-phase or activated sintering was rapid relative to coarsening. Undoped specimens heated at slower rates required longer soaking times to reach terminal densities, since the more extensive coarsening with slower rates lowered the driving forces for sintering during the soak. Slower heating rates for carbon-doped samples facilitated more extensive sintering at temperatures prior to the soak, shortening the required soaking period.

Hydrogen Treatments

Samples were soaked at various temperatures to permit time for gas-particle interactions under atmospheres of pure hydrogen, pure helium, or mixtures of hydrogen and helium. Samples were then heated up to 2230°C and soaked for 30 min under pure flowing helium.

Without soaking during the heating schedule, there was no difference in final density from the different atmospheres. Soaking at 1000°C also did not show noticeable changes, regardless of atmosphere. However, soaking at 1350°C generally showed a marked increase in density. Soaking periods in excess of 30 min did not appear to improve the final density. Hydrogen concentrations in excess of 40% during soaking at 1350°C resulted in lower densities than atmospheres with no hydrogen. Figure 6 shows that optimum densification resulted from sintering with a 10% H_2-90% He mixture during the soak at 1350°C. Figure 7 shows the dilatometry traces for specimens heat-treated in a selection of these atmospheric mixtures. Increasing H_2 concentration shifted the onset of sintering to lower temperatures. However, with increasing H_2 concentration, densification during continued heating and the soak at 2230°C was in fact less extensive.

Pre-sintering soaking at 1350°C in a H_2-He atmosphere successfully extracted some B_2O_3, presumably through the reaction $H_{2(g)} + B_2O_{3(l)} = H_2O_{(g)} + B_2O_{2(g)}$. Based on thermodynamic data, the equilibrium partial pressure of $H_2O_{(g)}$ is low, but not negligible (assuming $p_{H2} = 0.5$ atm, $p_{H2O} \approx 10^{-4}$ atm) at 1350°C. It is highly favorable for the H_2O vapor product (which was not drawn away by the flowing gas stream) to react with exposed B_4C: $B_4C + 7H_2O = 2B_2O_3 + 7H_2 +$

CO. However, the stoichiometry is such that a small amount of B_2O_3 is created relative to the H_2O available.

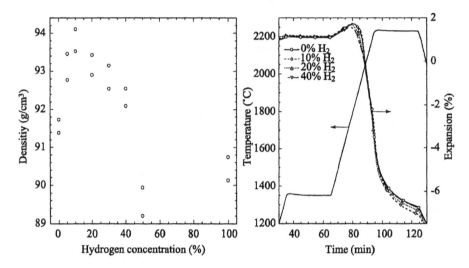

Figure 6 (left) Variation in post-sintered density with hydrogen gas concentration in a He/H_2 flowing gas mixture used during a 30 min soaking period at 1350°C.

Figure 7 (right) Contraction profiles from specimens sintered with a 30 min soak at 1350°C under different mixtures of flowing He/H_2. Heating rates before and after the soak were 30°C/min. The atmosphere was changed to pure He after the 30 min soak at 1350°C. Markers in this figure are used to identify traces, and do not represent data points.

From Figure 7, it is apparent that soaking in increasing concentrations of H_2 was increasingly effective at extracting B_2O_3 coatings from the particles, in turn permitting sintering to initiate at lower temperatures. The lingering presence of H_2 during heating to 2230°C proved disadvantageous. It is interpreted that hydrogen was absorbed interstitially into the B_4C particles, increasing the vapor pressure of B_4C (or its gaseous molecular subunits), in turn accelerating coarsening in this temperature range. The higher the H_2 concentration during the soak at 1350°C, the more remnant H_2 (not yet swept out by flowing He) present at temperatures at which coarsening was significant. Under rapid heating rates, minimum time for coarsening was permitted, which mitigated the effects of remnant H_2. When H_2 was purged from the 50-50 H_2-He mixture by rapidly flowing He for 2 h at 1350°C (Table III), the highest measured density, 94.7% of theoretical density, was obtained.

Adequate thermodynamic driving force for extraction of B_2O_3 particle coatings was not available when soaking in H_2/He at temperatures below 1350°C. Soaking in this atmosphere at 1600, 1750, and 1900°C permitted coarsening to occur by solution/precipitation in the B_2O_3 liquid phase while heating up to the soaking temperatures. Soaking at these temperatures served no purpose in B_2O_3 extraction, since it would have volatilized during heating prior to these temperatures. Further, the presence of H_2 during soaks at these temperatures would have accelerated coarsening by evaporation-condensation.

Table III: Boron carbide sintered under various heating schedules and atmospheres. The heating rate up to the soaking temperature was 30°C/min. After initial soaking, pure He was flowed into the furnace. Data represent single measurements.

Atmosphere	Soaking time (min) at 1350°C	Heating rate, 1350 to 2230°C, (°C/min)	Soaking time (min) at 2230°C	Final density (% of TD)
H_2(10%) + He(90%)	30	100	30	94.0
H_2(10%) + He(90%)	30	100	until 0% shrinkage	94.3
H_2(50%) + He(50%)	30	100	until 0% shrinkage	94.1
H_2(50%) + He(50%)	30, 100% He for 120 min	100	until 0% shrinkage	94.7

CONCLUSIONS AND FUTURE WORK

The density of undoped B_4C obtained after heating to 2250°C in He was 91.33 (% of TD). Recognizing evaporation-condensation coarsening in the range 2010-2140°C, rapid heating (100°C/min) to the same soaking temperature increased the sintered density to 92.76. A soaking period at 1350°C in a 50-50 H_2-He mixture, followed by a 2 h purge of H_2 (since it enhanced evaporation-condensation of B_4C at more elevated temperatures), resulted in a sintered density of 94.7%. Future work will involve improving green densities by improved starting particles and particle size distributions, use of pressing agents, and casting rather than pressing.

REFERENCES

[1]F. Thevenot, "Boron Carbide-A Comprehensive Review," *J. Euro. Ceram. Soc.*, 6 [4] 205-225 (1990).

[2]D. Emin, "Structure and Single-Phase Regime of Boron Carbide," *Phys. Rev. B*, 38 [9] 6041-6054, (1988).

[3]K. Niihara, A. Nakahira, and T. Hirai, "The Effect of Stoichiometric on Mechanical Properties of Boron Carbide," *J. Am. Ceram. Soc.*, 67 [1] c13-c14 (1984).

[4]G. de With, "High Temperature Fracture of Boron Carbide: Experiments and Simple Theoretical Model," *J. Mater. Sci.*, 19 [2] 457-466 (1984).

[5]G. A. Gogotsi, Y. L. Groushevsky, O. D. Dashevskaya, Y. G. Gogotsi, and V.A. Lavrenko, "Complex Investigation of Hot Pressed Boron Carbide," *J. Less-Common Met.*, **117** [4] 225-30 (1986).

[6]G. A. Gogotsi, Y. G. Gogotsi, and D. Y. Ostrovoj, "Mechanical Behavior of Hot Pressed Boron Carbide in Various Atmospheres," *J. Mater. Sci. Lett.*, **7** [8] 814-16 (1988).

[7]A. W. Weimer, "Thermochemistry and Kinetics," *Carbide, Nitride and Boride Materials Synthesis and Processing*, edited by A. W. Weimer, Chapman and Hall, New York, 79-113 (1997).

[8]R. Angers and M. Beauvy, "Hot Pressing of Boron Carbide," *Ceram. Int.*, **10** [2] 49-55 (1984).

[9]T. Vasilos and S. K. Dutta, "Low Temperature Hot Pressing of Boron Carbide and its Properties", *Am. Ceram. Soc. Bull.*, **53** [5] 453-54 (1974).

[10]K. A. Schwetz, W. Grellner, and A. Lipp, "Mechanical Properties of HIP-Treated Sintered Boron Carbide," *Inst. Phys. Conf. Ser. No. 75*, Chapter 5, Adam Hilger Ltd., Bristol, 413-26, (1986).

[11]S. L. Dole, S. Prochazka, and R. H. Doremus, "Microstructural Coarsening During Sintering of Boron Carbide," *J. Am. Ceram. Soc.*, **72** [6] 958-66 (1989).

[12]S. L. Dole, S. Prochazka, "Densification and Microstructure Development in Boron Carbide," *Ceram. Eng. Sci. Proc.*, **6** [7-8] 1151-160 (1985).

[13]S. Prochazka, S. L. Dole, and C. I. Hejna, "Abnormal Growth and Microcracking in Boron Carbide," *J. Am. Ceram. Soc.*, **68** [9] c235-c236 (1985).

[14]G. Agarwal, R. F. Speyer, and W. S. Hackenberger, "Microstructural Development of ZnO Using a Rate Controlled Sintering Dilatometer," *J. Mater, Res.*, **11** [3] 671-79 (1996).

[15]J. S. Reed, *Introduction to the Principles of Ceramic Processing*, 2nd Edition, John Wiley & Sons, New York, (1995).

[16]*The Columbia Encyclopedia*, 6th ed. www.bartleby.com/65/, Columbia University Press, New York, 2001.

[17]D. R. Lide, Editor in Chief, *CRC Handbook of Chemistry and Physics*, 74th Ed., CRC Press, Boca Raton, FL, 1993.

[18]M. Winter, *WebElements*, www.webelements.com.

[19]T. Y. Kosolapova, *Handbook of High Temperature Compounds: Properties, Production, Applications*, Hemisphere Pub. Co., New York, (1990).

[20]R. M German, *Sintering Theory and Practice*, John Wiley and Sons, New York, (1996).

[21]J. Beauvy, "System B-C, Solid State Phase Equilibria," *J. Les-Common Met.*, **90**, 169-75, (1983).

SPINEL ARMOR – CLEARLY THE WAY TO GO

Mark C. L. Patterson
Anthony A. DiGiovanni
Technology Assessment & Transfer Inc.,
133, Defense Highway, Suite 212
Annapolis, MD 21401

Gary Gilde
US Army Research Laboratory
AMSRL-WM-MC Building 4600
Aberdeen Proving Grounds
Aberdeen, MD. 21005

Don W. Roy
21210 North 132nd Drive
Sun City West, AZ. 85375

ABSTRACT

Spinel can be fabricated to show good transparency between approximately 0.25µm and 5.5µm and properties that make it attractive for a range of optical applications. Technology Assessment & Transfer Inc., in Maryland is presently scaling up the production of spinel for a wide range of applications. The present focus of this work is twofold; scaling up component fabrication capability for limited rate production of electro-optic components and increasing the size capability to achieve the fabrication of 0.5 inch thick spinel sections up to 22 inches in diameter for armor. While the properties of spinel are similar to other hard, optically transparent materials presently available (sapphire and ALON), spinel powder sources are readily available and it can be processed at lower temperatures and for shorter durations leading to a projected significant reduction in the fabrication cost. The present fabrication approach is by hot pressing (HP) followed by hot-isostatic pressing (HIP). High transparency can be achieved after HP alone but HIP is used to improve the final optical properties. Annealing is being investigated as an alternative approach to improving the optical properties.

This paper summarizes the present production capability and the projected scale up approach for armor. Recent ballistic comparisons between spinel, sapphire and ALON show a significant improvement for spinel and ALON over sapphire. An initial cost comparison indicates that spinel should be produced at a large cost reduction over sapphire or ALON. When fully developed, spinel should find extensive applications due to its performance, lower production cost, large fabrication scale capability and ease of finishing.

INTRODUCTION

Over the past 40 years spinel has been developed because of it's high hardness, high transparency in the MWIR and because of it's ease of fabrication[1]. It was not until the last decade however that spinel powders were produced, with sufficient purity and reactivity to make "good" quality transparent material with high transmission in relatively thick cross-sections. A number of attempts have been made over the past 10 years to commercialize spinel, Alpha Optical pursued HP of spinel with Coors until 1993 and RCS pursued rate controlled pressureless sintering until 1997, but progress was slow and spinel was only qualified for a few applications during this period. Technology Assessment and Transfer Inc., (TA&T) believe that there is a significant market for spinel both in the electro-optic (E/O) window and armor areas and has established a facility in Millersville MD, dedicated to the fabrication of spinel.

For the commercial production of spinel armor it is important that a stable and healthy business exists elsewhere. The rationale for this is that the armor business is sporadic, but the technology base needs to be stable and continuous to support and respond to the cyclic demands from armor. To this end TA&T have established a number of product and potential products for spinel E/O components to grow a stable and continuous base for spinel. One of the key properties driving the interest in spinel for E/O applications is the hardness combined with higher transmission in the MWIR as shown in Figure 1. This improved transmission is significant for targeting, tracking and data acquisition in the MWIR range and makes spinel the material of choice for future window and dome applications.

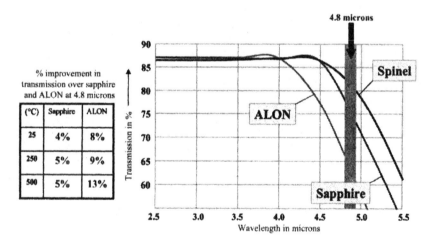

(°C)	Sapphire	ALON
25	4%	8%
250	5%	9%
500	5%	13%

% improvement in transmission over sapphire and ALON at 4.8 microns

Figure 1. Transmission spectrum at the MWIR absorption edge

The properties of polycrystalline ceramics can vary considerably based on the shape and size of the individual grains, the residual porosity and the nature of the grain boundaries. Transparent ceramics such as spinel typically contain very little residual porosity, but their mechanical properties can still be affected by large variations in the grain size and by the nature of the grain boundaries. Thus, the microstructure of spinel should be taken into consideration when discussing the mechanical properties. Likewise, the optical properties are also affected by the microstructure, and residual porosity at the triple points is thought to influence the scattering of light leading to a higher than usual haze.

Spinel crystal structure is cubic and optically isotropic; thus polycrystalline shapes may be fabricated without severe scattering problems inherent in polycrystalline non-cubic materials. In the microwave region the isotropy of spinel prevents localized absorption and heating that occurs in non-cubic materials because of differing grain boundary orientation and anisotropic dielectric loss index. Spinel undergoes no polymorphic transformations, so is free of problems due to thermally induced phase changes. Extensive programs were carried on in the 1980's to measure the properties of spinel as well as other candidate window materials, including sapphire, ALON and Yttria at Johns Hopkins University Applied Physics Laboratory[2], and Honeywell Systems Research Center[3].

In discussing material properties, it is important to consider the specific microstructures which have developed due to specific processing conditions. The flexural strength of spinel for instance, is strongly dependant on the final grain size and can be varied between approximately 350MPa and 100MPa, as the grain size is increased due to a final HIP temperature of 1700°C and 1900°C respectively. The 100MPa strengths have been attributed to abnormal grains that grow at these high temperatures to 2mm in size. It is not certain how the grain size impacts on the ballistic performance of a material but for the purpose of ballistic testing in this study, the grain size was large with a d_{50} of a few hundred microns. The work described in this paper summarizes three ballistic assessments, the first of which was generated from historic data for different armor configurations against a wide range of threats. The second and third sets of data were generated more recently under more controlled conditions and are considered comparable. In each case the threat is classified and has not been fully defined, but is an armored piercing projectile of approximately a .30 caliber.

With further development, spinel promises to be a low-cost, hard transparent material that will find widespread use across a range of E/O and transparent armor applications. The values reported in this paper for ballistic and cost projections have been normalized, but are the best projections at the time of publication.

EXPERIMENTAL APPROACH

The approach adopted by TA&T is to use commercially available powders, hot pressed at moderate temperatures to full transparency. Depending on the applications these dense part may be further HIPed to improve the optical transparency in the visible. The hot pressing is performed in graphite tooling using proprietary protocols. A 30 ton, 250 ton and 600 ton press are presently installed, although the 600 ton press is not scheduled to become operational until early in 2004. Pressure used for spinel range between approximately 1000psi and 5000psi although generally the pressure is varied between 2000psi and 4000psi. Figure 3 shows the scale of part achievable from each of these three presses is square inches, corresponding to 5", 15" and 22" square respectively. The large Birdsboro press has the capability of operating with a load of 1100 tons which would correspond to approximately 30" square as shown. Depending on the upcoming demand and technical development, TA&T may pursue this option in future years.

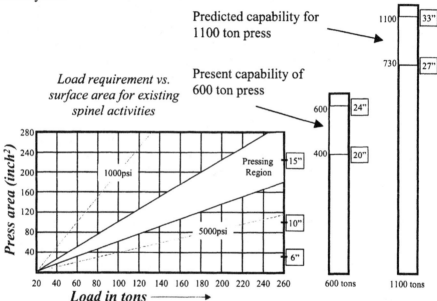

Figure 2. Projected pressing capability based on tonnage of presses installed.

A quality plan is presently in place, characterizing and documenting all variations from the raw powder, all tooling, operational conditions and a number of physical part properties such as post HP transmission. Selected samples are sectioned and polished to measure actual transmission, scatter and absorption values, and to measure grain size and distribution. For ballistic testing the spinel plates were surface ground to a 400 grit finish and backed with polycarbonate.

RESULTS

A large body of data has been collected on spinel plates 5" in diameter. The measurement of transmission before and after HIPing is routine, as is the measurement of refractive index uniformity following HIPing. The results described in this paper will summarize typical transmission data following HIPing, a collection of ballistic performance data against armored piercing (AP) threats and a comparison of fabrication costs for hard transparent materials.

Transmission

The transmission of selected parts has been measured following both HP and HIP procedures and the improvement brought about by both HIP and annealing, has been reported in previous publications [4],[5]. For a standard protocol, typical transmission values for 5" diameter, blank spinel parts are shown in Figure 3. These parts were pressed approximately 0.45" thick and reduced in thickness to 0.35" during the grinding and polishing operation. The transmission data is reported in three regions representing the visible, near and mid-wave infra-red. This data was collected by the University of Dayton Research Institute and corresponds to the center and edge of a 5" diameter plate and shows a slight improvement to the transmission at the edge with respect to the center of the plate.

Transmission with position – 0.35" thick

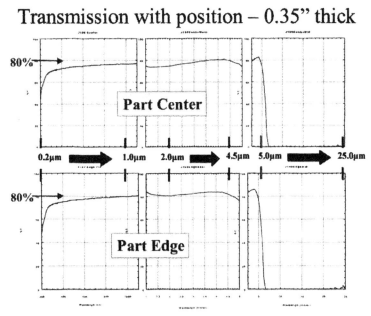

Figure 3. Transmission in the visible, near IR and MWIR for spinel plates 5" in diameter and 0.35" thick. Small special variations are observed between the edge and center of the part giving 80% and 78% transmission at 1μm respectively.

Ballistic Performance

The ballistic performance of spinel was measured for three different tests. In the first test (set #1), the performance of 4" square spinel plates was compared to historical data on glass, sapphire and ALON against similar projectiles. While the backing material, system overall thickness and projectile may have been slightly different for this evaluation, it shows very similar performance for sapphire, ALON and spinel and a significant improvement over the glass configuration. The data shown in Figure 4 is normalized for the projectile velocity and the aerial density of each of the armor systems. The second set of ballistic performance data (set #2) was collected more recently and compared spinel, glass and ALON with similar configuration against the same, unspecified AP threat. This data (set #2 in Figure 4) shows similar behavior for spinel and ALON and an approximate two to one advantage over glass systems.

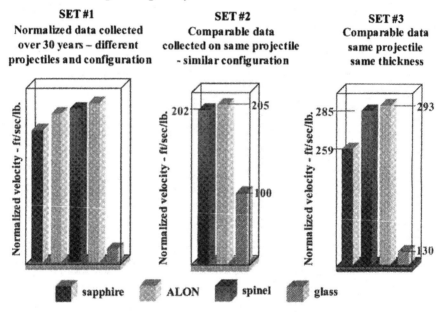

Figure 4. Ballistic performance for spinel, ALON, sapphire and glass against AP projectiles. Three individual assessments are shown

The final comparison (set #3) shows comparable armor systems evaluated against the same threat. The sapphire and spinel plates were all 5" in diameter, the ALON was 4" square and the ballistic glass was 12" square. All materials were of the same thickness. For each of these tests a minimum of 7 plates were shot for each material system. In this most recent comparable study (set #3) the spinel and ALON systems performed very similarly and showed an approximate 10% improvement over sapphire and a 100% improvement over ballistic glass.

Ceramic Armor and Armor Systems

Fabrication Cost Comparison

The true cost of producing materials and products is typically proprietary and often provides a competitive advantage to those who understand their competitors costs. The cost of producing spinel components will therefore only be discussed in a qualitative manner supported by circumstantial evidence.

Firstly, sinterable spinel powders are commercially available using solution precipitation, calcination methods and are typically around $100/kg. These powders are reactive enough for direct fabrication using conventional ceramic processing techniques such as pressureless sintering or HP. In comparison, ALON powders are not commercially available but require very high temperatures to form and additional milling to produce reactivity. High purity alumina powders are readily available and significantly lower cost than either spinel or ALON.

Full transparency can be achieved in spinel following HP at ~1600°C for short durations (< 1 day). Conversely, both sapphire and ALON require significantly higher fabrication temperatures for longer durations. Figure 5 shows a projected cost comparison for spinel, ALON, glass and sapphire based on cost information presently available. Set #1 compares production estimates for spinel based on HP followed by HIP. Generating and polishing costs for ALON and spinel are

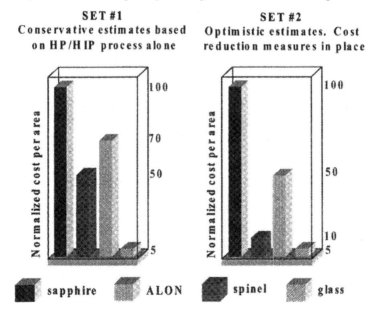

Figure 5. Preliminary cost comparison for sapphire, spinel, ALON and glass panels for present (set#1) and future (set #2) processes.

considered to be the same and slightly lower than sapphire due to it's higher hardness and issues surrounding crystal orientation which become important with shaped components. Set #2 provides optimistic estimates for spinel based on the assumption that significant improvements can be made during the scale up and development of a commercial operation. These predictions assume that sufficient transparency can be achieved for armor applications following HP followed by only an annealing cycle with a yield of 70%. They also assume a reduction in the polishing costs due to the application of a refractive index matched glass coating to the generated spinel surface[6]. Preliminary work has indicated that both of these assumptions are reasonable given time and a focused development effort.

CONCLUSIONS

A fabrication facility dedicated to the commercial production of spinel for E/O and armor applications is being established by TA&T. Transparent spinel plates up to 11" in diameter are presently produced and 22" diameter plates are scheduled to be produced in 2004. The ballistic performance of spinel has been evaluated against a range of AP threats and in each evaluation a significant improvement of at least 100% has been achieved for sapphire, spinel and ALON over ballistic glass systems. In the most recent evaluation using the same armor configuration (thickness, bond and backing materials) against the same threat spinel and ALON showed approximately 10% improvement over sapphire. The data has been classified and can be viewed by contacting Gary Gilde at the Army Research Laboratory in Aberdeen, MD. Spinel production costs promise to be significantly lower than either sapphire or ALON.

REFERENCES

[1] D.W.Roy, M.C.L.Patterson and G.Gilde, "Progress in the Development of Large Transparent Spinel Plates", 8th DoD EM Windows Symp., ASAFA, Colorado Springs, CO. 24th-27th April 2000

[2] M.E.Thomas, R.L.Joseph and W.J.Tropf, "Infrared Properties of Sapphire, Spinel and Yttria as a Function of Temperature", SPIE vol. 683, 1986.

[3] J.A.Cox, D.Greenlaw, G.Terry, K.McHenry and L.Fielder., "Infrared and Optical Transmitting Materials", SPIE vol. 683, 1986.

[4] M.C.L.Patterson, G.Gilde and D.W.Roy, "Fabrication of Thick Panels of Transparent Spinel" Inter. Symp. Proc. Optical Science & Technology. SPIE 46th Annual Meeting San Diego, CA. 29th July – 3rd August 2001.

[5] M.C.L.Patterson, G.Gilde and D.W.Roy, "An Investigation of the Transmission Properties and Ballistic Performance of Hot Pressed Spinel", Presented at the PAC RIM IV, Inter. Conf. Proceedings on Adv. Ceram. & Glasses. Wailea Maui, Hawaii, 4th-8th November 2001.

[6] P.McGuire, R.Gentilman, J.Askinazi and G.Gilde, "Glass Coated Sapphire for Low-Cost Transparent Armor", Proc. 9th DoD ElectroMagnetic Windows Symposium, 2002

Fracture Mechanism of Armor Ceramics and Composites

CONTROLLED EXPLOSIVE INDENTATION ON CERAMICS

Do Kyung Kim, Jong Ho Kim, Young-Gu Kim, and Chul-Seung Lee
Dept. of Materials Science and Engineering
Korea Advanced Institute of Science and Technology
Taejon, Korea

Dong-Teak Chung
School of Mechatronics
Korea University of Technology and Education
Chonan, Chungnam, Korea

Chang Wook Kim, Joon Hong Choi, and Soon-Nam Chang
Agency for Defense Development
Taejon, Korea

ABSTRACT

It is suggested that static mechanical properties and Hertzian indentation data allows to predict dynamic behavior of armor ceramics. Spherical indentation on several armor ceramics was conducted and yield strength and strain-hardening coefficient was determined for each specimens. In addition, controlled explosive indentation was conducted to evaluate the damage behavior of armor ceramics to simulate the impact loading. Simultaneously dynamic fracture of armor ceramics, based on static mechanical properties and Hertzian indentation data, was investigated by recently developed two dimensional explicit time integration finite element method. Simulation results show good agreement with experimental one. In conclusion, combination of Hertzian indentation and brittle fracture implemented finite element method would be a powerful tool for the prediction of dynamic fracture behavior of armor ceramics.

INTRODUCTION

Ceramic materials have good mechanical properties such as high strength in compressive loading, high elastic modulus, low density and high melting temperature. Some ceramic materials, such as alumina, silicon carbide and boron carbide, are primary candidates for advanced armor applications.[1, 2]

Even though the dynamic hardness of ceramics showed some relation with impact resistance,[3] it is not well understood about the dominant interaction of projectile or shaped charge jet with ceramics. The influence of mechanical properties of ceramic material on characteristics of armor ceramics is suggested. [4, 5]

Conventional fracture toughness, strength, and hardness do not provide enough information for selection of ceramics. When a projectile impacts on the ceramics, the stress and the damage distribution of ceramics are similar to that of Hertzian indentation. [6, 7] Hertzian indentation method has been widely used to evaluate ceramics. The Hertzian sphere indentation on ceramics can provide indentation stress-strain relation of ceramics over the wide range of strain. The compressive stress below the contact area reaches to tens of GPa even with the normal mechanical testing machine [8, 9], which is comparable to HEL(Hugoniot Elastic Limit) of the material. It is expected that Hertzian indentation would provide the information to predict dynamic fracture behavior.

For computer-based armor ceramics design, robust material modeling of brittle fracture behavior is required for finite element code. Most material models [10, 11, 12] are based on continuum damage theories in which the net effect of fracture is homogenized as a degradation of the strength of the materials and fragmentation size and density are represented by damage parameters. This continuum damage theory is not appropriate for brittle behavior because the discrete nature of crack is lost. Recently, in order to describe brittle fracture behaviors explicitly, including crack propagation, fragmentation, and node separation was proposed.[13] Also two dimensional explicit time integration finite element code of Lagrangian description was developed.[14] Simulation results of the code on dynamic fracture behavior of ceramics were reasonably comparable with experimental one.

In this study, the Hertzian indentation on armor ceramics with a spherical ball was conducted and indentation stress-strain relation was determined. Explosive detonator was used for dynamic indentation. FEM simulation with two dimensional explicit finite element methods associated with Hertzian indentation data, was conduced. The results from FEM simulation were compared with those of the controlled explosive indentation experiments.

EXPERIMENTAL AND SIMULATION
Hertzian indentation and mechanical properties

In this experiments, four armor ceramics, alumina(AD85), solid state sintered silicon carbide(S-SiC), hot-pressed silicon carbide(HP-SiC), and hot-pressed boron carbide(B_4C) were used. Specimens were cut into 6 x 8 x 35 mm and polished. Microstructure was examined by optical and scanning electron microscopy. Elastic modulus and Poisson's ratios were measured by pulse-echo method. Strength was measured by four-point flexure test and hardness and

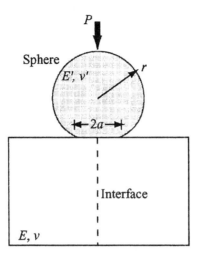

Figure 1. Hertzian contact of sphere on flat ceramic specimen. To reveal the subsurface damage, a special bonded-interface configuration was used. Indentation was made across the interface area.

toughness are measured by Vickers indentation. Figure 1 shows the schematic configuration of Hertzian indentation test. A spherical ball of radius r is pressed over the flat polished specimen. Beyond a critical load, either a Hertzian cone crack("brittle solid") or a subsurface deformation zone("plastic solid") initiates.[8] At normal load P, the contact radius a is given by

$$a^3 = \frac{4krP}{3E}, \quad \text{with } k = \frac{9}{16}\left[(1-\nu)^2 + (1-\nu)^2\frac{E}{E'}\right] \tag{1}$$

where prime notation denotes the indenter material. The contact radius a defines the spatial scale of the contact field. The mean contact pressure,

$$p_0 = P/\pi a^2 \tag{2}$$

defines the intensity of the contact field. From equation (1) and (2), the useful indentation stress strain relation is defined by

$$p_0 = \left(\frac{3E}{4\pi k}\right)\left(\frac{a}{r}\right) \tag{3}$$

which means a linear relationship between p_0, "indentation stress", and a/r, "indentation strain", leading to a procedure for obtaining basic stress-strain information. From the contact radius a and load P, the indentation stress and strain can be experimentally obtained. At sufficiently large contact area, some

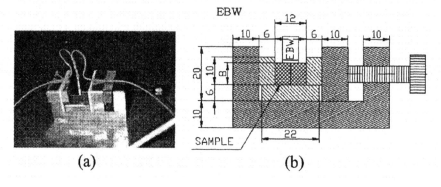

Figure 2. Experimental setup for the explosive indentation by small detonator.

form of plasticity generates before fracture, indicating a "brittle-ductile" transition. This transition from elastic to plastic contact can be conveniently demonstrated on indentation stress strain curve and the yield stress Y can be determined from transition point. Applied load was in the range of 500 and 2000 N. Indentation was also made on "bonded specimen" which was made and polished from two polished specimens bonded with the glue.[15] Spherical ball was loaded at the exact position of bonded interface and after detaching the sample using acetone, the subsurface damage mode was observed using optical microscopy with interference contrast after gold-coating treatment.

Controlled-explosive indentation

With a bonded-interface specimen sized 6 x 8 x 35 mm, the small explosive detonator with diameter of 5 mm was glued on the interface area. Light vice pressure was applied to avoid the shattering of the specimen during impact. Figure 2 shows experimental setup and geometry for the controlled explosive indentation by using a explosive detonator. The detonator is wrapped by stainless steel of 0.150 mm thickness and the stainless steel bottom disk with 5 mm diameter impacts on the surface of the ceramic specimen, when it is detonated, which simulates the dynamic indentation of armor ceramics. After indentation and detaching the specimen, the separated two side of the specimen were cleaned, gold-coated, and examined by the optical microscopy with interference contrast. SEM was used to observe the subsurface of damage zone with higher magnification.

Explosive indentation - numerical simulation

A two-dimensional explicit time integration finite element code of Lagrangian description was used for simulating the dynamic brittle fracture of ceramic. In this

Ceramic Armor and Armor Systems

code, when the traction is greater than the critical spall strength, the tensile crack can grow. This behavior was described by

$$\sigma_{eff} = \sqrt{\sigma^2 + \beta_\tau \tau} \geq \sigma_{in}$$

$$\sigma_{in} = \sigma_{fr} e^{t/t_c} /(e^{t/t_c} - 1)$$

(4)

where σ_{eff} is effective traction, σ, τ, and β_τ are normal and tangential tractions and shear stress factor, and σ_{fr}, τ_c, and τ are quasi-static spall strength, characteristic time and pulse duration time. These cracks can branch, coalesce, and lead to the fragments eventually. Also under dynamic compressive load, plastic deformation of ceramic was described as like the granular materials, with Y_f being commuted material strength, α and β being model parameters, ε_f being fracture strain, P being applied pressure, s being degradation factor, and Y_c being compressive inert yield strength. In the process of numerical computation, new surface and fragments are created by the results of node separation and element eroding. Careful treatments for multibody and self-contact situations are required. Used computation code includes contact situation [16] and more accurate calculation is achieved by adopting and implementing defense node algorithm. [17]

In numerical calculation of explosive indentation, it is assumed that stainless steel disk having 0.150 mm of thickness, 5 mm of diameter, 1 km/s of initial impact velocity, 10 mm of radius of curvature. Ceramic specimens placed on steel fixture. Axisymmetric model was used and the input parameters came from static properties and indentation data, as listed in Table I. Considering the observation of experimental, the half of flexural strength was used for split stress in ceramics exhibiting mostly intergranular fracture, i.e., HP-SiC. From the results of simulation, the development of crack and spalling was observed as time scale and final evolution of damage was compared with experimental results.

Table I. Physical properties of armor ceramics. (abbrev: processing method, manufacturer; AD85: sintered Al2O3, Coors; S-SiC: sintered, Carborumdum; HP-SiC: hot-pressed, Norton; B4C: hot-pressed, Cercom)

sample	d (g/cm^3)	ν	E (GPa)	σ (MPa)	H (GPa)	T (MPa$\sqrt{m}$)	Y (GPa)
AD85	3.439	0.23	236	266	9.2	3.23	6.5
HP-SiC	3.217	0.174	442	525	19.6	3.75	9.93
S-SiC	3.166	0.168	440	553	29.1	2.46	9.5
B4C	2.503	0.167	456	390	27.3	3.66	10.84

(d denotes density, ν Poisson's ratio, E elastic modulus, σ flexural strength, H hardness, T toughness, and Y yield strength)

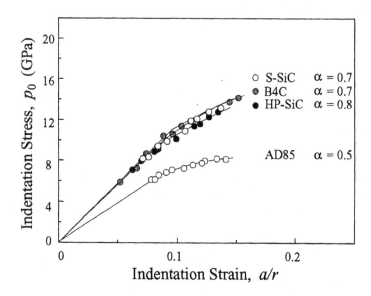

Figure 3. Hertzian indentation stress-strain curves are plotted for each specimen. Data points are from the experiments and solid curves are fitting of simple relation with yield strength and strain-hardening coefficient.

RESULT AND DISCUSSION

Static Sphere Indentation (Hertzian indentation)

Table I shows the physical properties of four ceramics. HP-SiC, S-SiC, and B_4C shows high hardness, modulus, and yield strength value. Strength shows near 500 MPa in case of HP-SiC and S-SiC, and relatively low value in case of AD85. Toughness values of four ceramics show almost same value. These measured static data are adopted for input parameters for numerical simulation of explosive indentation. Figure 3 represents the indentation stress-strain relation of ceramics. In the curve, S-SiC, B_4C, and HP-SiC shows high yield strength (Y) and α. If α equals to zero, it shows fully plastic behavior and if α equals to 1, it is considered as a fully elastic materials. This yield strength value is considered to be the limit of elastic regime during impact loading. It is considered that the indentation stress-strain relationship is useful to predict impact damage behavior of ceramics.

Explosive indentation – experimental

Figure 4 shows side views of explosive-indented ceramics. All samples show circular damage on top surface having same dimension of bottom of explosive. Below the contact, massive radial and lateral cracks were observed. In Fig. 4(a), subsurface damage of AD85 was shown. Near contact region contains severe damage similar damage pattern of static Hertzian indentation. Also lateral and radial cracks were developed by overlap of transmitting and reflecting shock waves are observed. Intensively damaged regions marked with dotted line in the side view are apparently distinguishable from outer region. Quasi-plastic deformation zone was developed in the strong shear-compression region below the contact. In contrast with AD85, high yield strength materials, S-SiC and B₄C have shown the spall-dominant damage patterns near top and bottom surface. It is considered that high yield strength affects on the minimum quasi-plasiticity. High flexural strength S-SiC shows low spalling but relatively low flexural strength

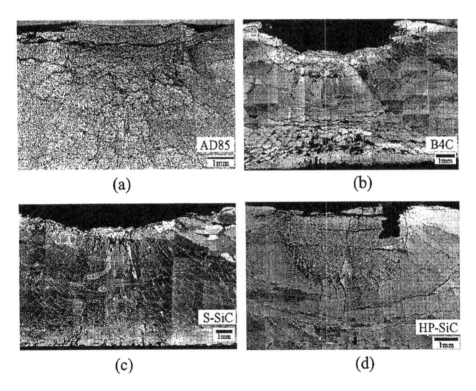

(a)

(b)

(c)

(d)

Figure 4. Side views of explosive-indented (a) AD85, (b) B₄C, (c) S-SiC, and (d) HP-SiC ceramics showing quasi-plastic and fragmentation damage zone. Optical microscopy with interference illumination highlights the detail contrast of damage zone.

B$_4$C shows large spalling patterns. These two materials could be highly resistive to impact loading. HP-SiC and S-SiC have similar physical properties, however these two ceramics shows quite different damage mode. S-SiC shows spalling dominant damage and HP-SiC shows quasi-plastic damage. This particular feature of damage has its origin in damage mode of ceramic during impact loading. SEM observation of damaged zone reveals that HP-SiC has intergranular fracture mode but S-SiC has transgranular fracture mode. During impact loading on ceramic, intergranular fracture enhances crack propagation. In polycrystalline ceramics, Lawn[18] has documented that the generic fracture mechanical model of the microfracture evolution within the subsurface damage zone during a full damage is associated with the activation of discrete "shear faults", from which microcracks initiate. The overall features of microfracture in damage zone of explosive-indented specimen were almost same as that of statically indented one.

FEM simulation of explosive indentation

Figure 5 shows FEM simulation result as time history, with AD85 specimen. In this figure, left side reveals pressure contour during indentation and right side shows crack development. At t = 0.4 µs, concentrated compression below contact exceeds elastic limit and plastic damaged occurs in near contact region. Half circle-shaped pressure inside the specimen propagates through thickness direction and leaves damage behind. Cone crack and quasi-plastic-like damage were observed at t = 0.8 µs. As time elapsed, crack propagated along lateral and radial direction and intensive cracks appear below contact. The reflection of shock wave leaves d amage n ear b ottom free s urface. Finally t he s triker p late b ounces from specimen surface and partial spalling of specimen occurs. Simulation have performed up to t = 100 µs after impact, but the damage of all the ceramics were completed at about t = 20 µs.

Comparison of simulation and experimental results are shown in Fig. 6. The simulation results of AD85 (Fig. 6a) shows a good agreement with experimental. Partial spalling and dispersed cracking were observed both in experimental and simulation. C rack p atterns a re n ot e xactly s ame b ut o verall f eatures o f damage show a general well agreement between simulation and experimental results. As mentioned in explosive indentation section, S-SiC and B$_4$C show spalling dominant response to impact loading. Simulation results of B$_4$C and S-SiC show the same dynamic fracture behavior as experimental observed one. As seen in figure 6(b) and figure 6(d), S-SiC and B$_4$C show no severe microcracking damage inside the specimen, but B$_4$C exhibits relatively large spalling both near contact and bottom surface. Figure 6(c) shows the result of HP-SiC. Distributed cracking in simulation is well matched with quasi-plastic zone in experiment. The smallest spalling is also same f eature which have been o bserved in controlled explosive experiment.

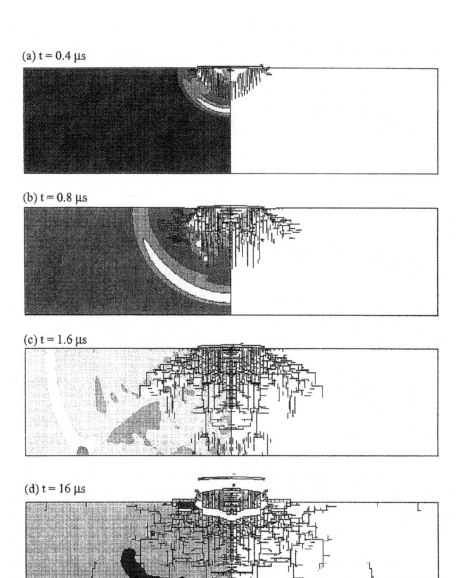

Figure 5. Simulated internal crack patterns of AD85, overlapped with pressure contour (left half) at (a) t = 0.4 μs, (b) t = 0.8 μs, (c) t = 1.6 μs, and (d) t = 16 μs after explosive indentation. Lighter area in pressure contour represents higher pressure.

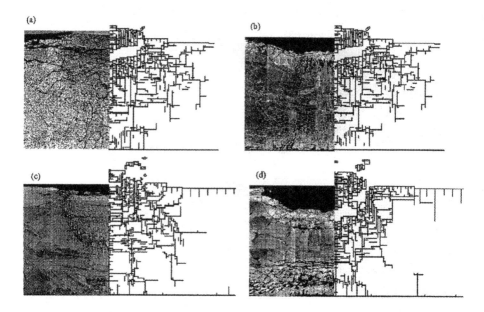

Figure 6. Experimental subsurface damages (left half) are compared with simulation results of crack pattern (right half) for each ceramics, (a) AD85, (b) S-SiC, (c) HP-SiC, and (b) B$_4$C. Crack patterns are simulation results at t = 20 μs after explosive indentation.

CONCLUSION

Dynamic indentation damage of four typical armor ceramics was examined by FEM simulation and the controlled explosive indentation experiment. Using parameters from the static Hertzian indentation and conventional mechanical properties, the dynamic impact damage behavior of armor ceramics was successfully simulated. Simulation results were comparably well matched with the result of dynamic indentation experiment. It was shown that the damage behavior of ceramics under dynamic impact could be related with the conventional static physical properties, including yield strength.

It can be suggested that the static Hertzian indentation has an advantage of cost and convenience to get the yield strength of ceramics. And the controlled explosive indentation test would be an alternative useful methodology for the quick examination of dynamic damage evolution of armor ceramics with a minor materials parameter, such as microstructural variables.

Ceramic Armor and Armor Systems

REFERENCES

[1] C. Donaldson, "The Development of a Theory for Design of Lightweight Armor," Technical Report AFFDL-TR-77-114, Aeronautical Research Association of Princeton, Inc.

[2] William A. Gooch Jr. "An Overview of Ceramic Armor Applications," pp. 3-21 in Ceramics Transactions, vol. 134, Ceramic Armor Materials by Design, edited by J.W. McCauley, et. al., The American Ceramic Society, 2002.

[3] D.B. Marshall and A.G. Evans "Measurement of Dynamic Hardness by Controlled Sharp-Projectile Impact," J. Am. Ceram. Soc. 66 [8] 580–585 (1983).

[4] D.K. Kim, C.-S. Lee, C.W. Kim, and S.N. Chang, "Indentation Damage Behavior of Armor Ceramics," pp. 429–39 in Ceramics Transactions, vol. 134, Ceramic Armor Materials by Design, edited by J.W. McCauley, et. al., The American Ceramic Society, 2002.

[5] D.K. Kim, C.-S. Lee, Y.-G. Kim, C.W. Kim, and S.N. Chang, "Dynamic Indentation Damage of Ceramics," pp. 261–68 in Ceramics Transactions, vol. 134, Ceramic Armor Materials by Design, edited by J.W. McCauley, et. al., The American Ceramic Society, 2002.

[6] A.G. Evans and T.R. Wilshaw, "Quasi-Static Solid Particle Damage in Brittle Solids-I. Observation, Analysis and Implications," Acta Metall. 24, 939–56 (1976).

[7] A.G. Evans and T.R. Wilshaw, "Dynamic Solid Particle Damage in Brittle Materials: An Appraisal," J. Mater. Sci. 12, 97–166 (1977).

[8] B.R. Lawn, "Indentation of Ceramics with Spheres : A Century After Hertz," J. Am. Ceram. Soc. 81 [8] 1977–94 (1998).

[9] B.R. Lawn and T.R. Wilshaw, "Indentation Fracture : Principles and Applications," J. Mater. Sci. 10 [6] 1049–81 (1975).

[10] L.M. Taylor, E.P. Chen, and J.S. Kuszmaul, "Microcrack-Induced Damage Accumulation in Brittle Rock under Dynamic Loading," Comp. Meth. App. Mech. 55, 301–320 (1986).

[11] A.M. Rajendra, "Modeling the Impact Behavior Od AD85 Ceramic under Multi-axial Loading," Int. J. Impact Eng. 15 [6] 749–68 (1994).

[12] A.M. Rajendran and D.J. Grove, "Effect of Pulverized Material Strength on Penetration Resistance of Ceramic Targets," Proceedings of the 1995 APS Shock Compression Conference, 1995.

[13] G.T. Camacho and M. Ortiz, "Computational Modeling of Impact Damage in Brittle Materials," Int. J. Solids. Structures, 33, 2899–938 (1996).

[14] D.T. Chung, C. Hwang, S.I. Oh, and Y.H. Yoo, "A Study on the Dynamic Brittle Fracture Simulation," The 4th Int. Symp. on Impact Eng., Kumamoto, Japan, July 16-18, 2001.

[15]. B.R. Lawn, N.P. Padture, H. Chai, and F. Guiberteau, "Making Ceramics Ductile," Science, 263, 1114–16 (1994).

[16] M. Oldenburg and L. Nilsson, "The Position Code Algorithm for Contact Searching," *Int. J. Numer. Meth. Eng.* **37**, 359–386 (1994)

[17]. Z.H. Zhong, *Finite Element Procedure for Contact-Impact Problems,* Oxford University Press, 1993.

[18] B.R. Lawn, N.P. Padture, F. Guiberteau, and H. Cai, "A Model for Microcrack Initiation and Propagation Beneath Hertzian Contacts in Polycrystalline Ceramics," *Acta Metall. Mater.* **42** [5] 1683–93 (1994).

EVIDENCE OF DUCTILE (ALUMINA) AND BRITTLE (BORON CARBIDE) RESPONSE OF CERAMICS UNDER SHOCK WAVE LOADING.

E.B. Zaretsky and V.E. Paris
Department of Mechanical Engineering
Ben-Gurion University of the Negev
P.O.B. 653, Beer-Sheva 84105,
Israel

G.I. Kanel
Institute for High Energy Densities
Russian Academy of Sciences
Izhorskaya 13/19, Moscow, 125412
Russia

A.S. Savinykh
Institute of Problems of Chemical Physics,
Russian Academy of Sciences,
Chernogolovka, Moscow region, 142432,
Russia

ABSTRACT

An experimental technique capable of revealing the mode of inelastic, ductile or brittle, behavior of materials under shock-wave compression is presented. The technique utilizes an essential difference between the dependencies of brittle failure threshold and ductile yield stress on the lateral pressure: the brittle failure strength increases rapidly with the pressure whereas the ductile yield stress is much less pressure-sensitive. A controlled lateral pressure (pre-stressing) was produced by installing a shrink-fit steel sleeve on the lateral surface of the disk-like sample. Assuming that the tested ceramic sample obeys either the Von Mises criterion of ductile yielding or the Griffith's criterion of brittle failure, it was expected that the increment of the Hugoniot Elastic Limit caused by the pre-stressing should differ by a factor of about 2.5 for these two cases. The results of the tests performed unambiguously exhibit the ductile response of the alumina ceramic whereas the response of the boron carbide ceramic is certainly brittle.

INTRODUCTION

It is known that cracking of brittle solids caused by an axial compression may be suppressed by applying a hydrostatic pressure. Heard and Cline[1] found that the static response of brittle BeO and AlN ceramics became ductile when the axial loading of the ceramic sample was accompanied by lateral confinement with pressure of about 0.5 GPa; on the contrary the alumina samples remained brittle even at confining pressures up to 1.25 GPa. The physical mechanisms governing the ductile or brittle inelastic material straining are different and, respectively, they should be described by different constitutive models and criteria. That is why the conditions of the brittle–ductile transition are crucial for predicting the compressive response of brittle solids. However, the capabilities of quasi-static high-pressure tests are too limited to provide the transition parameters for high-strength ceramic materials.

Planar impact experiment is a way to obtain quantitative information about resistance of materials to inelastic deformation over a practically unlimited pressure range at sub-microsecond load durations. The plate impact tests are widely used for studying the strength properties of different materials, including high-strength ceramics, and, as well, for calibrating their constitutive models. Unfortunately, until now, no way to detect the mode of the response of the brittle materials above their elastic limits was found, and the question of whether the observed elastic limit corresponds to the yield strength or to the compressive failure threshold is still open. In the plate impact tests the shock-wave loading occurs under conditions of uniaxial compression whereas in most of other tests the uniaxial stress condition is utilized. The strength data of different kinds of the tests may be compared and unified by determining the threshold values for some equivalent stress in various strength criteria. However, until the mode of the response, ductile or brittle, of the tested material is unknown, the conciliation of the results of the tests of different kinds is hardly possible. It is generally accepted to characterize the transition of ductile material from elastic to inelastic deformation using the von Mises or Tresca yield criteria. In the case of the uniaxial strain loading conditions the criteria coincide and yield the relation between the stress σ_1 acting in the impact direction and the transversal stress σ_2:

$$\sigma_1 - \sigma_2 = Y_{duct},\tag{1}$$

where the yield strength Y_{duct} is the threshold value for the equivalent stress. Accounting in that for the uniaxial strain loading conditions $\sigma_2 = v\sigma_1/(1-v)$, where v is the Poisson's ratio, the experimentally observed stress σ_1 corresponding to the onset of the ductile yielding (the Hugoniot Elastic Limit, HEL) is equal to

$$\sigma_1 = \sigma_{HEL} = Y_{duct}\frac{(1-v)}{(1-2v)},\tag{2}$$

and the yield strength determined from the σ_{HEL} value is equal to

$$Y_{duct} = \sigma_{HEL} \frac{(1 - 2v)}{(1 - v)}.$$

(3)

Analyzing the shock-wave experiments with different ceramics Rosenberg[2, 3] arrive at a conclusion that for interpreting the HEL of brittle materials it is quite reasonable to use the Griffith's criterion of brittle failure instead of the Tresca (von Mises) criterion of ductile yielding. The Griffith's criterion is based on the assumption that the compressive failure starts when the tensile stress acting on the largest crack of the most vulnerable orientation exceeds some critical value causing the crack extension. In the uniaxial strain case of planar impact loading the use of the Griffith's criterion gives

$$(\sigma_1 - \sigma_2)^2 = Y_{brit}(\sigma_1 + \sigma_2),$$

(1a)

The threshold value Y_{brit} of the equivalent failure stress calculated with aid of (1a) from the same experimentally measured value of σ_{HEL} is

$$Y_{brit} = \sigma_{HEL} \frac{(1 - 2v)^2}{1 - v}.$$

(3a)

It can be seen from comparison of Eqs. (3) and (3a) that using the measured σ_{HEL} for calculating the equivalent stress Y yields the value of Y corresponding to the start of the brittle failure by the factor of $(1 - 2v)$ lower than that corresponding to the onset of the ductile material yielding.

It seems attractive to expand the capabilities of planar impact experiments so as to determine the mode of the inelastic behavior of brittle materials. Heard and Cline[1] found that below the brittle-to-ductile transition the compressive strength of brittle ceramics increases rapidly with the confining pressure whereas above the transition the strength is much less sensitive to the pressure. The different sensitivity to the variations of confining pressure may be used for detecting the mode of inelastic response although in the case of uniaxial shock wave loading the possibility of such variations is limited: the transversal stress σ_2 is nonzero *a priori*; for σ_1 less than σ_{HEL} it is equal to $\sigma_2 = \sigma_1 v/(1 - v)$. In particular, for alumina ceramics the transversal stress at the HEL is varied from 1.5 to 3.5 GPa depending on the material porosity, composition, and grain size. Chen and Ravichandran[4, 5] in their experiments with split Hopkinson pressure bar technique provided a controlled radial pre-stressing by a shrink-fit metal sleeve on the lateral surface of the cylindrical ceramic specimen. This technique of varying the transversal stress was used in our recent planar impact tests with alumina.[6] Present work describes further development of the method and new experimental data.

DEPENDENCIES OF THE HUGONIOT ELASTIC LIMIT ON THE RADIAL STRESS FOR DUCTILE AND BRITTLE MATERIALS

Applying an additional compressive stress π (in what follows, this stress will be called pre-stressing) in the direction normal to the principal loading stress σ_1 will change the expressions (1) and (1a). In this case the transversal stress σ_2 should be substituted by $\sigma_2 + \pi$. The substitution yields, respectively, the Hugoniot Elastic Limits of ductile and brittle solids

$$\sigma_{HEL}^{duct} = Y\left(1 + \frac{\pi}{Y}\right)\frac{(1-v)}{(1-2v)}, \tag{4}$$

$$\sigma_{HEL}^{brit} = \frac{Y}{2}\frac{1-v}{(1-2v)^2}\left[2(1-2v)\left(\frac{\pi}{Y}\right) + 1 + \sqrt{1 + 8(1-2v)(1-v)\left(\frac{\pi}{Y}\right)}\right]. \tag{4a}$$

Differentiating Eqs. (4) and (4a) with respect π gives, in the case of $\pi/Y \ll 1$,

$$\frac{d\sigma_{HEL}^{duct}}{d\pi} = \gamma^{duct} = \frac{(1-v)}{(1-2v)} \tag{5}$$

$$\frac{d\sigma_{HEL}^{brit}}{d\pi} = \gamma^{brit} = \frac{(1-v)(3-2v)}{(1-2v)} \tag{5a}$$

As apparent from (5) and (5a) the sensitivity of σ_{HEL} to the pre-stressing is in $(3-2v)$ times higher when the material response is brittle. In particular, the ratio $\gamma^{brit}/\gamma^{duct}$ is about 2.5 for aluminas ($v \approx 0.23$) and about 2.7 for boron carbide ($v \approx 0.17$). Comparing the measured σ_{HEL} values for the stress-free and the pre-stressed samples allows one to verify whether the criterion of brittle failure or the criterion of ductile yielding is applicable and what is the mode of the inelastic deformation that takes place in the investigated material under shock-wave loading.

MATERIALS

The experiments were performed with hot pressed alumina (96% purity) and boron carbide (97% purity) samples (Microceramica Ltd., Haifa, Israel). The samples were precisely (with 2-μm tolerance) cut disks of 25-mm diameter and 5-mm thickness. The properties of four tested alumina samples and seven boron carbide samples are given in Tab. I. The density of the samples was determined by liquid displacement in distilled water. The longitudinal C_l and the shear C_s sound speeds were measured by ultrasonic pulse-echo technique using 5-MHz probes of 12.5-mm diameter.

The densities and the sound speeds were found practically the same, $\rho_0 = 3.745 \pm 0.005$ g/cm^3, $C_l = 10.03 \pm 0.01$ km/s and $C_s = 5.93 \pm 0.02$ km/s, for all the alumina samples. The corresponding Poisson's ratio is $v = 0.231$ and the Young's modulus is $E = 324$ GPa.

Table I. Average mechanical properties of studied Al$_2$O$_3$ and B$_4$C samples.

Material	C_l, km/s	C_s, km/s	C_b, km/s	ρ_0, g/cm^3	Poisson's ratio v	E, GPa
Al$_2$O$_3$ average	10.03 ($\pm$0.01)	5.93 ($\pm$0.02)	7,33 ($\pm$0.06)	3.745 ($\pm$0.005)	0.231 ($\pm$0.002)	324 ($\pm$2)
B$_4$C average	13.89 ($\pm$0.25)	8.75 ($\pm$0.05)	9.52 ($\pm$0.36)	2.493 ($\pm$0.009)	0.17 ($\pm$0.02)	447 ($\pm$10)
BC-A	13.91	8.83	9.46	2.493	0.163	452
BC-B	14.12	8.78	9.82	2.499	0.185	456
BC-C	14.15	8.77	9.88	2.494	0.188	456
BC-D	13.79	8.65	9.51	2.501	0.176	440
BC-E	14.09	8.74	9.83	2.497	0.187	453
BC-F	13.48	8.75	8.92	2.474	0.136	430
BC-G	13.67	8.75	9.21	2.495	0.153	440

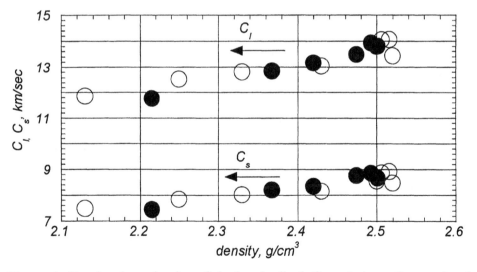

Figure 1. Density dependencies of the longitudinal C_l and shear C_s speeds of sound in boron carbide. Filled symbols show the data of present work, open symbols are the data of other studies.[7-10]

The densities of boron carbide samples were found highly scattered within the range of 2.215 ± 0.005 to 2.500 ± 0.005 g/cm^3 (the ultimate density of B$_4$C is 2.52 g/cm^3). The density dependencies of both C_l and C_s (see Fig. 1) were found very close to those of other authors.[7 - 10] The average grain size in the studied B$_4$C samples was found to be ranged from 15 to 20 μm, close to the size of the B$_4$C grains in the samples studied by Gust and Royce.[7] The Young's moduli of these samples varied between 290 GPa for $\rho_0 = 2.21$ g/cm^3 and 460 GPa at $\rho_0 = 2.50$ g/cm^3. Seven boron carbide samples having densities close to the ultimate one, Tab. I, had been chosen for this study.

EXPERIMENTAL RESULTS AND CONCLUSIONS

The technique suggested by Chen and Ravichandran [4, 5] for variable pre-stressing of the disk-like samples was found suitable for planar impact experiments of the present work. The pre-stressing of the ceramic samples was produced by shrink-fit steel rings of outer diameter 45 mm machined from rods of normalized 4340 steel with $E = 200$ GPa and $\nu = 0.28$. The inside diameter of the rings was smaller than the diameter of the samples by $\delta = 0.1\pm0.005$ mm. Prior to the insertion of the ceramic samples, the rings were heated up to 600°C to ensure the ring expansion sufficient for easy insertion of the cold sample. The ring shrinkage accompanying cooling of the assembly provided the radial pre-stressing of the sample. The pre-stress π produced by the ring was calculated using known solution of an axis-symmetric boundary value problem[10]. For purely elastic deformation of the ring the solution is

$$\pi = \frac{\delta}{2R_c\left[\dfrac{1-\nu_c}{E_c} + \dfrac{R_s^2 + R_c^2 + \nu_s\left(R_s^2 - R_c^2\right)}{E_s\left(R_s^2 - R_c^2\right)}\right]}, \qquad (6)$$

where R_c is the ceramic sample radius, R_s is the steel ring outer radius, E_c, ν_c, and E_s, ν_s are the Young's moduli and Poisson's ratios of the sample and the ring materials, respectively. For the case of the elastic-perfect plastic behavior of the ring material the pre-stressing is given by the expression

$$\pi = \sigma_Y\left[\ln(R/R_c) + \frac{R_s^2 - R^2}{2R_s^2}\right], \qquad (7)$$

where R is the radius of the boundary between elastically and plastically deformed regions of the ring. The boundary radius may be found from the equation

$$\left[\left(1-2v_s\right)\left(1+v_s\right)\frac{\sigma_Y}{E_s}-\left(1-v_c\right)\frac{\sigma_Y}{E_c}\right]\left[\ln\left(\frac{R_c}{R}\right)-\frac{R_s^2-R^2}{2R_s^2}\right]+$$

$$+\frac{\left(1-v_s^2\right)}{E_s}\sigma_Y\frac{R^2}{R_c^2}=\frac{\delta}{2R_c}. \tag{8}$$

Mechanical properties of the ring material were determined using the witness samples machined from the same, as the ring, steel rod and subjected to the heating procedure simultaneously with the rings. The measured values of the yield and ultimate tensile stress of the steel are equal to $\sigma_Y = 0.78\pm0.02$ GPa and $\sigma_{UTS} = 0.92\pm0.02$ GPa, respectively, so the description of 4340 steel as an elastic–perfect plastic material seems reasonable. For the misfit $\delta = 0.1$ mm the state of steel ring falls just a little outside the elastic range and confining stress is equal to $\pi = 0.3$ GPa for the alumina samples and 0.32 GPa for the boron carbide samples.

Table II. Properties of free and pre-stressed B₄C samples[a].

Sample, density	Thickness, mm	C_l, km/s	C_s, km/s	Poisson's ratio v
BC-B, 2.499 g/cm³	5.082 - 5.094	14.12 - 13.71	8.78 - 8.86	0.185 - 0.141
BC-C, 2.494 g/cm³	5.067 - 5.076	14.15 - 13.48	8.77 - 8.75	0.188 - 0.136
BC-E, 2.497 g/cm³	5.081 - 5.086	14.09 - 13.87	8.74 - 8.77	0.187 - 0.167
BC-G, 2.495 g/cm³	5.073 - 5.077	13.67 - 13.93	8.75 - 8.83	0.152 - 0.164

[a] The left values are of the free samples.

The thicknesses and the sound speeds measurements were repeated with the boron carbide samples after completing the pre-stressing. These data, together with the corresponding initial parameters of stress-free samples are given in Tab. II and Fig. 2. What is striking in these data that the sample thickness increments are substantially higher than 1 to 1.2μm expected for the 5-mm sample having $E \approx 450$ GPa and $v = 0.16$. The pre-stressing produces practically negligible changes of the shear sound velocities C_s whereas variations of C_l correlate with the variations of the sample thickness. The observed effects of pre-stressing on both the sample thickness and the sound speed may be considered as a manifestation of the dilatancy phenomenon: the radial compression opens a part of the micro-cracks existing in the virgin material. Accounting in that all the samples are of the same porosity, it is plausible to assume that the observed effect of the pre-stressing on the properties variations is related to the micro-cracks orientation and shape.

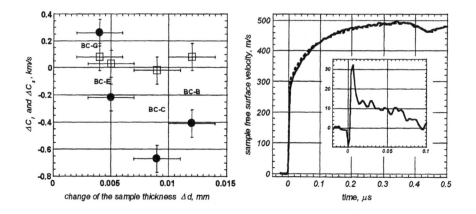

Figure 2. Changes of the longitudinal C_l (filled circles) and shear C_s (squares) speeds of sound vs. changes of the sample thicknesses caused by the pre-stressing of boron carbide samples.

Figure 3. Free surface velocity data obtained with free of stress (solid lines) and pre-stressed (dashed lines) samples of alumina (a). Insert shows the difference between two average curves.

The shock wave loading of alumina ceramic samples was carried out by impact of 1-mm copper flyer plate launched by a 57-mm gas gun facility with the velocity of 500±10 m/s. The planar impact loading of first three, Tab. I, boron carbide samples was carried out by 1-mm tungsten flyer plates at the impact velocity of 580±11 m/s. The impactor-sample misalignment was controlled in each gun experiment by three flush pins and did not exceed 0.5 mrad. The stronger shots with boron carbide samples (BC-D to BC-G samples of Tab. I) were performed with use of explosive facility accelerating aluminum flyer plates of 2-mm thick up to impact velocity of 1900±50 m/s. In the experiments with alumina ceramics the samples free surface velocity histories, $w(t)$, were recorded with a VISAR[10]. In the experiments with the boron carbide samples the velocity histories were recorded for interface between the sample and the PMMA window of 5.88-mm thickness.

Figure 3 summarizes the results of experiments with pre-stressed and free alumina samples. Each shown free surface velocity profile was obtained by averaging of the profiles of two shots. The transition from elastic to inelastic response occurs at the free surface velocity $w_{HEL} = 285$ m/s that corresponds to $\sigma_{HEL} = \rho_0 C_l w_{HEL}/2 = 5.35$ GPa. The ramped velocity growth behind the elastic wave front is associated with strain hardening and stress relaxation effects. The insert shows the difference $w_c - w_u$ of free surface velocities of confined, w_c, and unconfined, w_u, ceramic samples. The oscillation of this value at very early times

Ceramic Armor and Armor Systems

is obviously the result of insufficient temporal resolution of the measurements (2 ns) and non-reproducible dispersion of the shock front.

Assuming that the inelastic response of the alumina is ductile the confining pressure $\pi = 0.3$ GPa should result, Eq. (4), in the increase of the HEL equal to $\Delta\sigma_{HEL} = 0.43$ GPa. The latter corresponds to the increase of the sample surface velocity by $\Delta w_{HEL} = 23$ m/s. In the case of the brittle inelastic response the Griffith's failure criterion results in HEL stress and particle velocity increments that are 2.5 times greater, so the expected free surface velocity increment should be close to 60 m/s. The measured difference between the average $w(t)$ data for pre-stressed and free ceramics for the first 50 ns is 15 to 10 m/s.

The results of the experiments with boron carbide samples are shown in Fig. 4. Like in the experiments of Kipp and Grady [8] the waveforms are oscillated. The oscillations, which are greater at higher stress, are, obviously, caused by strongly heterogeneous inelastic deformation with relatively large distances between the heterogeneities (localized shears or cracks). The elastic precursor wave in boron carbide has some more complicated structure than in the alumina. In most of the shots the elastic compression wave consists of a shock discontinuity of less than 1-ns rise time followed by a gradual, some 10-%, velocity increase during 25–50ns.

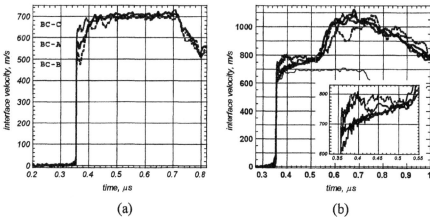

(a)   (b)

Figure 4. Velocity histories of sample-PMMA interfaces obtained with free (solid lines) and pre-stressed (dashed lines) samples of boron carbide in weak impact (a) and strong impact (b) experiments. The hair line (b) shows the velocity profile of weak impact experiment with BC-C sample. The enlarged domain of interest is shown in the insert.

The rise time of the second, inelastic compression wave varied from 40 ns to 70 ns. The intermediate part of the waveform between these two compression waves is changed dramatically by the pre-stressing. For the free samples this part of the waveform presents a gradual growth similar to that observed for alumina. In the pre-stressed samples, a plateau, or even some decreasing the velocity, and, hence, the compressive stress replaces the gradual increase.

Since the transitions from the elastic shock discontinuity to the intermediate part of the wave profile are not regular and reproducible, we supposed that this feature is caused heterogeneous deformation of the ceramic. So, we define the HEL as the intersection of the upward extrapolation of the elastic shock discontinuity with the extrapolation of the intermediate part of the waveform to the left. Thus, as apparent from the comparison of Fig. 4(a) with Fig. 4(b), the Hugoniot Elastic Limit was exceeded in none of the low-impact experiments. The differences between the low-velocity profiles are caused by specific for each sample, but still elastic, process of the pores (micro-cracks) closing. In the strong impact experiments, Fig. 4(b), the HEL of the samples was certainly exceeded and, so, the values of the B_4C/PMMA interface velocities equal to $u_{ifHEL} = 700\pm10$ m/s and $u_{ifHEL} = 775\pm10$ m/s may be related to elastic-inelastic transitions in the free of stress and the pre-stressed samples, respectively. Using Barker and Hollenbach [13] data on shock Hugoniot of PMMA yields for the HEL of the free and pre-stressed B_4C samples the values $\sigma_{HEL} = 13.5\pm0.15$ GPa and $\sigma_{HEL} = 15.1\pm0.2$ GPa, respectively. So, the pre-stressing of the boron carbide samples by 0.32 GPa resulted in the average increase of the HEL of about $\Delta\sigma_{HEL} \approx 1.6\pm0.35$ GPa. Equation 4(a) yields the HEL increment the value of $\Delta\sigma_{HEL} \approx 1.06\pm0.03$ GPa for brittle response whereas the expected increment for ductile response should be, in accordance with Eq. (4), of 2.7 times less. The discrepancy, 1.6 GPa vs. 1.06 GPa, seems due to the use of the free sample failure threshold Y_{brit} derived from the Griffith's criterion (1a) which is known[2] yields the underestimated values of Y_{brit}. Like in the case of alumina, the difference between the "free" and the "pre-stressed" velocity profiles decreases to zero after a lapse of about 150 ns.

The results unambiguously show the ductile response of alumina and the brittle response of boron carbide under conditions of one-dimensional shock compression. In the case of the alumina the lateral stress at the HEL in those experiments was varied between 1.6 (free samples) and 1.9 GPa (pre-stressed samples). That means that in alumina the transition from brittle to ductile response occurs between the confining pressure 1.25 GPa, at which, as was shown by Heard and Cline[1], alumina remains brittle, and 1.6 GPa. In the case of the boron carbide samples the highest confining pressure, corresponding to the highest detected u_{ifHEL} value and equal to 2.97 GPa, is apparently insufficient for brittle to ductile transition.

ACKNOWLEDGMENT

This work was partially supported by the Basic Research Found of the Israeli Ministry of Defense within the framework of Project No. 84297-101.

Ceramic Armor and Armor Systems

REFERENCES

[1] H.C. Heard and C.F. Cline, J. Mat. Sci., 15 1889 (1980).

[2] Z. Rosenberg, J. Appl. Phys., 74 752 (1993).

[3] Z. Rosenberg, J. Appl. Phys., 76 1543 (1994).

[4] W. Chen, and G. Ravichandran, J. Am. Ceram. Soc. 79, 579 (1996)

[5] W. Chen and G. Ravichandran, Internat. J. of Fracture, 101 141 (2000).

[6] E.B. Zaretsky and G.I. Kanel. Appl. Phys. Letters 81(7) 1192-1194 (2002).

[7] W.H. Gust and E.B. Royce, J. Appl. Phys. 42, 276–295 (1971).

[8] M.E. Kipp and D.E. Grady, in: Shock Compression of Condensed Matter—1989 (eds. S.C.Schmidt, J.N. Johnson, and L.W. Davison) North-Holland, Amsterdam, pp. 377-380 (1990).

[9] D.E. Grady, in: High-Pressure Science and Technology—1993 (eds. S.C. Schmidt, J. W. Shaner, G.A. Samara, and M. Ross) American Institute of Physics, New York, pp. 741–744 (1994)

[10] N.S. Brar, Z. Rosenberg, and S.J. Bless, in: Shock Compression of Condensed Matter—1991 (eds. S.C. Schmidt, R.D. Dick, J.W. Forbes, and D.G. Tasker), Elsevier Science Publishers, B.V., North-Holland, Amsterdam, pp. 467-470 (1992).

[11] R.D. Cook and W.C. Young, Advanced Mechanics of Materials, Macmillan Publishing Company, NY, 1985

[12] L.M. Barker and R.E. Hollenbach, Journ. Appl. Phys. 43, 4669 (1972)

[13] L.M. Barker and R.E. Hollenbach, Journ. Appl. Phys. 41, 4208 (1970)

DYNAMIC FAILURE AND FRACTURE OF THE TiC-STEEL COMPOSITE.

E. Zaretsky   N. Frage and M.P. Dariel
Department of Mechanical Engineering   Department of Materials Engineering
 Ben-Gurion University of the Negev, P.O. Box 653, Beer-Sheva 84105, Israel

ABSTRACT

The dynamic response of a titanium carbide (TiC)- steel ceramic-metal composites (cermets), prepared by pressureless infiltration of TiC ceramic preforms by molten steel, was studied in planar impact experiments, using copper impactors with velocities in the 80 to 450 m/sec range. The effect of the microstructure of the metallic component on the Hugoniot Elastic Limit (HEL) and on the spall strength of the cermet samples was studied by use of the VISAR (Velocity Interferometer System for Any Reflector) for measuring the velocity of the free surface of the samples. The study revealed that the confining stress, developed in the carbide component of the cermet by the strain-hardened steel matrix under tension, affects strongly the dynamic response of the cermet. The value of the confining stress that was estimated on the base of two types of spall measurements correlates with the increase of the HEL with respect to that of the cermet containing a stress-free ceramic sub-matrix. This correlation leads to two definite conclusions:

i. The TiC component of the cermet fails in compression like a brittle material, while the inelastic response of the cermet is similar to that of a metal.

ii. The dynamic response of the cermet may be controlled by choosing an appropriate thermal treatment.

INTRODUCTION

Ceramic-metal composites (cermets) have a potential for light armor applications[1]. Design of the cermet armor is possible when data on its dynamic response are available. At present, such data are scarce and non-systematic. The main efforts of the studies of such composites were focused on the influence of the morphology of ceramic inclusions on the dynamic properties of Metal Matrix Composites (MMCs) [2-8], or cermets.[7 - 11] The influence of the initial state of the metal matrix on the dynamic response of cermets has not been addressed in those investigations. Such influence was studied for first time by the authors of the present work in the TiC-carbon steel cermets[12], containing the steel matrix in normalized, quenched or tempered at 250°C states. Later[13], the scope of the study was widened by adding more tempering regimes. The present work continues the study of the dynamic response of the TiC-steel cermets.

MATERIALS AND EXPERIMENTAL

The titanium carbide-steel cermets consisted of two interpenetrating and interconnected, ceramic and metal, networks. The TiC-steel cermet samples (the comprehensive description of their fabrication may be found elsewhere[13]) differed by the state of the metal matrix, the volume fraction of which was between 29 to 35% and the chemical composition close to that of 1080 steel. The samples that had undergone no thermal treatment, except that associated with their fabrication, are referred henceforth as (a). The (a)-samples quenched after austenizing, are referred as samples (q). And, finally, the (q)-samples tempered after the quenching are referred as samples (t_T), where T=150, 200, 250, 300, 350 and 400°C is the temperature of the one-hour-long tempering treatment. Samples of a similar size (20-mm diameter disks of 3-4-mm thickness and base planes parallel to within 0.005-mm) were cut from a rod of commercial, normalized 1080 steel that had undergone the same heat treatments. The high-density (96%) TiC samples produced of 2 μm-fine powder were used for determining the properties of the ceramic component in the absence of the metallic network.

The microstructure analysis of the polished cermet samples ("JEOl-35" scanning electron microscope) show that the thermal treatment changes the structure of the metallic matrix from clearly pearlitic, characteristic for (a)-samples, to fine-plate martensite in (q)-samples, and, finally to the differently (depending on the tempering temperature) developed bainite in (t_T)-samples. The longitudinal C_l and shear C_s velocities of propagation of acoustic waves were determined by pulse-echo ultrasonic technique and were found almost independent of the applied thermal treatment both for the cermet and the steel samples. The average static properties are summarized in Table I.

Two types of the planar impact experiments (25-mm gas gun, hollow aluminum projectiles equipped with copper impactors of 2-mm thickness) were used for the study of the TiC-steel cermets and the ceramic TiC samples: (i) "strong" impact experiments (H-shots), with the impactor velocity of about 400 m/sec, providing an impact-generated stress amplitude higher than the HEL (σ_{HEL}) of the sample, and (ii) "weak" L-shots providing the stress amplitude below the HEL.[13] The steel samples were studied by H-experiments only. The velocities w of the free surface of the samples were continuously monitored by VISAR.[14] The values of σ_{HEL} and of the dynamic tensile strength σ_{spall}^{H} in the H-shots were determined from the HEL velocity amplitude, w_{HEL}, and the spall

118

velocity pull-back, Δw_{spall}, of the free surface velocity profiles, using the expressions $\sigma_{HEL} = 0.5\rho_0 C_l w_{HEL}$ and $\sigma^H_{spall} = 0.5\rho_0 C_0 \Delta w_{spall}$, where the bulk sound velocity is $C_0 = [C_l^2 - (4/3)C_s^2]^{0.5}$. For calculating the spall strength in the L-shots (pure elastic loading) the expression $\sigma^{BH}_{spall} = 0.5\rho_0 C_l \Delta w_{spall}$ was used.

TABLE I. Average static properties of the studied materials.

Materials and treatment	Density, g/cm³	C_l, km/sec	C_s, km/sec	Poisson's ratio	G, GPa	E, GPa
TiC	4.77(±0.03)[a]	10.09(±0.08)	6.10(±0.03)	0.216[b]	178[b]	430[b]
Cermets	5.58(±0.04)	8.62(±0.08)	5.12(±0.05)	0.228	150	367
Steels	7.84(±0.01)	5.94(±0.02)	3.24(±0.02)	0.293	80	206

[a] Standard error of the value is given in parentheses.
[b] The values are calculated on the base of average data.

RESULTS

Hardness measurements (Buehler micro-hardness tester, Vickers indenter, 20 N load) confirmed the formation of a hard component at the outcome of quenching. Hardness data are related to the quasi-static inelastic behavior (yield strength) of the tested material. In contrast to the elastic moduli, the hardness of both the cermet and steel is strongly affected by the heat treatment (Tab. II). The similar trend followed by the hardness values of the cermet and steel samples, as the result of the heat treatment, allows us to assume that it is the steel sub-matrix that controls the inelastic response of the cermets. The results of the study of the dynamic response of the cermets, described below, support this assumption.

Table II. Vickers hardness numbers of differently treated samples

	as is (a)	quenched (q)	t_{200}	t_{250}	t_{300}	t_{350}	t_{400}
Cermets	1080	1750	1960	1500	1350	1200	1380
Steels	285	850	800	743	-	-	546

Typical profiles of the free surface velocity w of the studied samples are shown in Fig. 1. Notice that the origin on the sample free surface velocity (y-axis) for the sake of clarity has been shifted upwards for the samples that had undergone the heat treatments (q) and were tempered at 250 °C (t_{250}). The dimensional-less time is $\tau = tC_l/\delta$, where δ is the sample thickness and t is the time elapsed after impact. The Tab. III gives the averaged values of the σ_{HEL}, σ^L_{spall}, and σ^H_{spall}.

A comparison of the velocity profiles, shown in Fig. 1, for the above-HEL shots of steel samples with those for the cermets that have undergone the same heat treatment, reveals a definite correlation. Quenching and tempering the samples increase the yield strength of both the steel and the cermet, with respect to the untreated samples. The influence of the heat treatment on the yield strength of the material may be estimated quantitatively by approximating the compressive part of the free surface velocity profile by a centered simple wave[15] with the slope of the centered characteristics equal to the Lagrangian velocity $a(u) = h/t(u)$ of the propagation of the perturbation, u, and assuming for the particle velocity u one half of the free surface velocity, w. Integrating the equations of mass and momentum conservation along the characteristics yields the parametric form of the compressive curve of the studied material $\sigma(\varepsilon) = \sigma[\varepsilon(w)]^{13}$, while the values of the stress deviator $s(\varepsilon)$ and the flow stress $Y(\varepsilon)$

$$s(\varepsilon) = \frac{2}{3}Y(\varepsilon) = \sigma(\varepsilon) - p(\varepsilon),$$ (1)

may be found from the pressure-volume dependence derived from the linear Hugoniot, $D = C_0 + Su$ of the material: $p(V) = C_0^2(V_0 - V)/[V_0 - S(V_0 - V)]^2$.

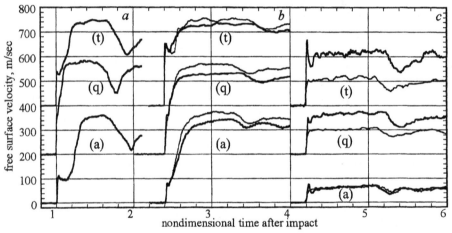

Figure 1. VISAR records of the sample free surface velocity profiles obtained in the shots with steel, a, and cermet, b and c, samples after different heat treatments (shown near the profiles). Pairs of profiles obtained after shots with cermet samples (b and c) are shown to illustrate the typical scatter of the dynamic properties.

The dependence $Y(\varepsilon)$, derived from the free surface velocity profiles of the steel and the cermet samples are shown in Fig. 2. The same value of $S = 1.35$, close to that of TiC ($S = 1.35$) and Armco-Fe ($S = 1.33$) was used for all materials.[16] The heat treatment produces a substantial increase of the flow stress, Y, both in the steel and in the cermets (Fig. 2), whereas the results of tempering the steel and the cermet differ: while some partial softening of the quenched steel takes place after tempering at 250°C, the yield stress of the adequately tempered cermet is higher than that of the quenched one.

The flow stress of quenched cermets and of those tempered at 150 and 200°C are very close and characterized by strong strain dependence, close to that of quenched steel. The strain dependence of the flow stress in the cermets tempered at 250, 300 and 400°C resembles that of the steel tempered at 250°C, and, finally, the effects of the treatment temperature on the strain hardening moduli $dY/d\varepsilon$ of the cermet and steel are nearly identical, Fig. 4, and show clearly that the inelastic deformation of the cermet is controlled by the metallic matrix.

The influence of the tempering temperature on the cermet spall strength, σ^L_{spall}, measured after the L-experiments, was found similar, with a distinct maximum for tempering at 250°C, to that observed for the HEL, Fig. 5. The values of the σ^H_{spall} corresponding to the above-HEL shots do not display any significant temperature dependence.

TABLE III. Average dynamic properties of the studied materials.

Materials and treatment	σ^{BH}_{spall}, GPa	σ^{AH}_{spall}, GPa	σ_{HEL}, GPa
TiC	0.24(±0.02)[a]	-	5.27(±0.09)[a]
Cermet (a)	0.82(±0.02)	0.65(±0.05)[a]	1.67(±0.04)
Cermet (q)	1.25(±0.04)	0.74(±0.05)	3.47(±0.02)
Cermet (t_{150})	1.52(±0.08)	0.82(±0.08)	4.00(±0.20)
Cermet (t_{200})	1.55(±0.10)	0.90(±0.10)	4.15(±0.15)
Cermet (t_{250})	1.60(±0.10)	0.68(±0.05)	4.60(±0.20)
Cermet (t_{300})	1.40(±0.05)	0.78(±0.08)	4.10(±0.20)
Cermet (t_{350})	1.30(±0.05)	0. 70(±0.10)	3.78(±0.01)
Cermet (t_{400})	1.25(±0.01)	0.73(±0.05)	2.80(±0.10)
Steel (a)	-	2.54[b]	2.24[b]
Steel (q)	-	2.37[b]	3.28[b]
Steel (t_{250})	-	2.49[b]	2.91[b]

[a] Standard error of the value is given in parentheses.
[b] Obtained from the single impact experiment.

Since no data on the dynamic response of the TiC matrix were available, planar impact experiments were performed on dense TiC samples. The HEL of the dense TiC sample was determined in H-shot and was found equal to $\sigma_{HEL} =$ 5.27 GPa, while determination of the spall strength of this sample appeared impossible: the dense TiC sample was completely destroyed in compression above HEL and the spall signal was absent at the corresponding velocity profile. An L-impact experiment yields $\sigma_{spall} = 0.24$ GPa. for dense TiC.

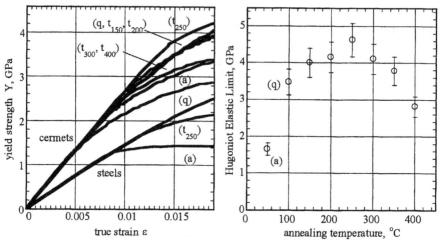

Figure 2. Dynamic yield behavior of the steel and the cermet samples.

Figure 3. Average values of Hugoniot Elastic Limit of cermet samples that had undergone a different heat treatment. The annealing temperatures 50 and 100 °C are arbitrarily assigned to the "as is" (a) and quenched (q) samples, respectively. The error bars correspond to approximately ±10% scatter of the measured σ_{HEL} values.

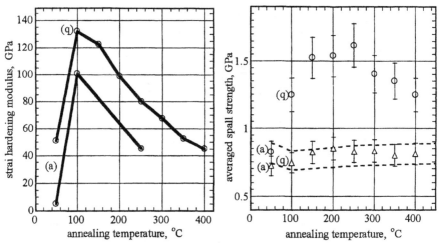

Figure 4. Effect of the tempering temperature on the strain hardening moduli $dY / d\varepsilon$ of the cermet (upper line) and the steel (bottom line).

Figure 5. Average values of σ_{spall}^H (triangles) and σ_{spall}^L (circles) of cermet samples that had undergone a different heat treatment. Dotted lines correspond to 29% (bottom) and 35% (upper) of the spall strength of steel samples after tempering at different temperatures. The spall strength of (a)-sample was used for extrapolation to the elevated annealing temperatures.

Ceramic Armor and Armor Systems

DISCUSSION

As just mentioned, any compressive deformation above HEL leads to the complete destruction of the ceramic. The relation between the σ_{HEL} and the yield stress, Y, corresponding to the failure of the brittle material in compression, was derived by Z. Rosenberg[17, 18] from Griffith's yield criterion

$$\sigma_{HEL} = \frac{1-\nu}{(1-2\nu)^2} Y, \tag{2}$$

where Poisson's ratio, $\nu = 0.216$ for dense TiC, (Tab. I). Using the measured value $\sigma_{HEL}^{TiC} = 5.27$ GPa, Eq. (2) gives for the yield stress of the dense TiC, $Y^{TiC} = 2.17$ GPa. According to Griffith's biaxial-stress yield criterion, $Y = 8\sigma_0$, where σ_0 is the material tensile strength under uniaxial stress tension. Assuming $\sigma_{spall}^{TiC} = \sigma_0^{TiC}$ we obtain $Y^{TiC}/\sigma_{spall}^{TiC} = 8$, while from the experimental data, $Y^{TiC}/\sigma_{spall}^{TiC} = 2.17/0.24 = 9$. The agreement allows concluding that the dense TiC is a typical brittle ceramic in which shock compression above σ_{HEL}^{TiC} results in its complete disintegration.

Complete destruction of the TiC matrix is also observed in above-HEL impact experiments on cermets. As apparent from Fig. 5, the spall strength of the cermets in H-shots is almost independent of the state of the metallic matrix and lies between 0.71 and 0.85 GPa. It is plausible to assume that during the tensile path of the H-shots, the effective cross-section of the cermet is actually the cross section of the metallic matrix (29 to 35% after metallographic analysis), since the ceramic component has been destroyed in the course of the compressive path of the loading cycle. This assumption is justified by comparing the σ_{spall}^{H} values of all the studied cermets with $0.29\,\sigma_{spall}^{steel}$ to $0.35\,\sigma_{spall}^{steel}$, Fig. 5. In the L-shots, (the circles in Fig. 5), the influence of the thermal treatment on the σ_{spall}^{L} is evident. The increase may be attributed to some excess compressive stress $p^* = \Delta\sigma_{spall} = \sigma_{spall}^{L} - \sigma_{spall}^{H}$, generated within a ceramic particle by the steel envelope. This compression prevents the premature opening of the cracks in the ceramic particle and, thus, maintains the integrity of the sample cross section. It was shown[13] that the cermet can be described as a 3-dimensional quasi-periodic structure of identical cells containing a ceramic particle surrounded by a steel envelope. This approach is similar to the "method of cells" used by Aboudi [20, 21] for calculating the elastic moduli of MMCs. The 30/70 steel-to-TiC particles volume ratio results in an average ratio of the steel envelope thickness, δ, to the particle radius, R, of about $\delta/R \approx 0.2$. The compressive stress, p^*, acting on the ceramic particle implies the presence of a tensile stress σ_t within the steel envelope. For the upper limit of such tensile stress the value[19] $\sigma_{tR} = p*R/2\delta$ may be assumed.[13] According to Fig. 5, the highest excess stress $p*$ is observed in the samples that were tempered at 250°C. In these samples the tensile stress $\sigma_{tR} = p*R/2\delta$ may reach 2 – 2.2 GPa. The compressive curves of the quenched and tempered steel samples, shown in Fig. 2, seem to validate the ability of the steel envelope to withstand such tension. It is relevant to point out that some of the cermet samples were partially cracked just after the quenching but no new cracks appeared after tempering. The compressive path of the below-HEL impact loading does not seem to change substantially the pre-stressed state of the steel matrix. The fracture of the sample starts when the excess compressive stress $p*$ is cancelled in tension.

The presence of the excess stress $p*$ may explain the substantial changes of the σ_{HEL} of the cermet samples induced by the heat treatment. As shown, the yield behavior of dense TiC follows obeys the Griffith's criterion. The cermets were manufactured from similar TiC preforms and differed only in the state of their metal component. The results obtained for the steel samples indicate that the effect of the thermal treatment differed substantially from that observed in the cermets. It is plausible to attribute the different response to the mechanical interaction between the steel and TiC matrices. In the presence of the excess pressure, $p*$, the stress σ_1 and σ_2 in Griffith's criterion

$$\left(\sigma_1 - \sigma_2\right)^2 = Y\left(\sigma_1 + \sigma_2\right) \tag{3}$$

has to be replaced by $\sigma_1 + p*$ and $\sigma_2 + p*$, respectively. Solving (3) with the assumption that the yield stress, Y, and Poisson's ratio, ν, of the ceramic are independent of the pressure $p*$ and accounting for the 1-D strain conditions, $\sigma_2 = \sigma_1 \nu/(1-\nu)$, yields

$$\sigma_1 = \sigma_{HEL} = \frac{1}{2}Y\frac{(1-\nu)}{(1-2\nu)^2}\left[1 + 2\frac{p*}{Y}\frac{(1-2\nu)\nu}{(1-\nu)} + \sqrt{1 + 8\frac{p*}{Y}(1-2\nu)(1-\nu)}\right]. \tag{4}$$

In the case of a small compressive stress $p* \ll Y$, the pressure derivative of σ_{HEL} is:

$$\left.\frac{d\sigma_{HEL}}{dp*}\right|_{p*\to 0} = \frac{\left(2 - \nu + \nu^2\right)}{(1-2\nu)}. \tag{5}$$

Poisson's ratio of the studied cermets varies from $v = 0.21$ to $v = 0.23$ (see Tab. I), thus

$$\left.\frac{d\sigma_{HEL}}{dp*}\right|_{p*\to 0} = 3.16 \div 3.38 . \tag{6}$$

Assuming that the inelastic response of the cermet is ductile and is governed by the von Mises (Tresca) criterion,

$$\left(\sigma_1 - \sigma_2\right) = Y , \tag{7}$$

yields for the pressure derivative of σ_{HEL}

$$\left.\frac{d\sigma_{HEL}}{dp*}\right|_{p*\to 0} = \frac{v}{(1-2v)} . \tag{8}$$

For values of the Poisson ratio in the same, $v = 0.21$ to $v = 0.23$, range,

$$\left.\frac{d\sigma_{HEL}}{dp*}\right|_{p*\to 0} = 0.36 \div 0.43 . \tag{9}$$

Comparing the numerical values of (6) and (9), we have to conclude that the brittle character of the inelastic response of a material causes the stronger dependence of the material HEL on the excess compressive stress. This statement is in agreement with the experimental results of Heard and Cline [22] for BeO and AlN. The observed strong effect of the state of the metal sub-matrix on σ_{HEL} of the cermet allows us to assume that the yield behavior of the ceramic component stays brittle within the whole range of the excess pressure generated by the metal component. The σ_{HEL} of cermets that had undergone different heat-treatments, as a function of the $p* = \Delta\sigma_{spall}$, along with the least square linear approximation of the data are shown in Fig. 6. The agreement between the approximating equation $\sigma_{HEL} = \sigma^0_{HEL} + kp* = 1.64 + 3.27p*$ and the prediction, Eq. (6), seems adequate.

The Griffith failure criterion (3) is a threshold criterion: in the principal space plane $\left(\sigma_1, \sigma_2 = \sigma_3\right)$ the corresponding failure surface is described by the expression $\sigma_1 = \sigma_2 + 0.5\left(Y + \sqrt{Y^2 + 8Y\sigma_2}\right)$ derived from Eq. (3). Two such surfaces, the bottom one corresponds to the cermet with unconfined ceramic particles, the upper one corresponds to the confinement pressure $p* = 0.95$ GPa,

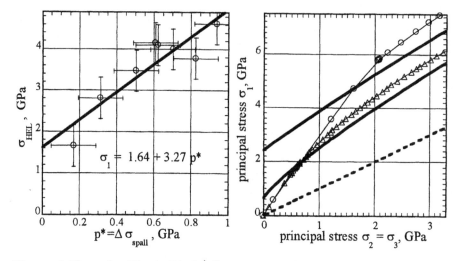

Figure 6. Hugoniot Elastic Limit of cermet samples as a function of the excess pressure p^* (circles), and square least approximation of the pressure dependence of HEL according Eq. (5).

Figure 7. Failure surfaces of (a) (bottom solid line) and (t_{250}) (upper solid line) cermet samples shown in the principal stress space together with the corresponding compressive curves, triangles and circles, respectively. The dashed line corresponds to a pure hydrostatic response. $Y = 1.64 \cdot (1 - 2v)^2 / (1 - v) = 0.66 \, GPa$, see Fig. 6.

are shown in Fig. 7. As soon as the brittle body being compressed elastically fails, the following inelastic compression is governed by the inter-particle friction, [23] corresponding, in the principal stress plane, to the origin-centered straight lines lying below the failure surface. If the failed material is unable to support any shear stress, its response should be purely hydrostatic [23], as shown by the dashed line in Fig. 7. The triangles and the circles in the Fig. 7 correspond to the impact-generated trajectories of the (a)- and (t_{250})-samples, respectively, on the principal stress plane obtained from the corresponding compressive curves, Fig. 2, within the assumption that $\sigma_1 = \sigma(\varepsilon)$ and $\sigma_2 = \sigma_1 - Y(\varepsilon)$, see Eq. (1). It is apparent from Fig. 7 that although the cermets fail in a brittle manner, the presence of the metal matrix contributes to support the relatively high level of the shear stress.

CONCLUSIONS

The results obtained in planar impact experiments lead us to conclude that the dynamic response of TiC-steel ceramic-metal composites is governed by the mechanical interaction between the ceramic and metallic sub-matrices. A model based on describing the cermet as a three-dimensional periodical arrangement of unit cells, consisting of ceramic spheroid particles encased in a steel shell under

Ceramic Armor and Armor Systems

tension, allows estimating the interaction and following its evolution as a function of the applied heat treatment. The obtained experimental data allow to construct, at the principal stress plane, the failure surfaces of differently treated cermets and to recognize that the presence of the metal matrix within the cermets, even in the post-failure state, permits maintaining the shear stress at the relatively high level.

REFERENCES

[1] S.J. Bless, D.L. Jurick and S.P. Timothy, in International Conference on High Strain Rate Phenomena in Materiasl, San Diego 1990, Eds. M.A. Meyers, L.E. Murr and K.P. Staudhammer, Elsevier, New-York, 1991, p.187.

[2] G. T. Gray III, in High-Pressure Science and Technology-1993, Eds. S.C. Schmidt, J.W. Shaner, G.A. Samara and M. Ross, AIP Conference Proceedings 309, New-York, 1994, p.1161.

[3] J. N. Johnson, R. S. Hixson, and G. T. Gray III, J.Appl.Phys., 76, 5615, (1994).

[4] R. S. Hixson, J. N. Johnson, G. T. Gray III, and J.D. Price, in Shock Compression of Condensed Matter-1995, Eds. S.C. Schmidt, and W.C. Tao, AIP Conference Proceedings 370, New-York, 1996, p.555.

[5] G. T. Gray III, R. S. Hixson, J. N. Johnson, in Shock Compression of Condensed Matter-1995, Eds. S.C. Schmidt, and W.C. Tao, AIP Conference Proceedings 370, New-York, 1996, p.547.

[6] R.U.Vaidya, S.G. Song, A.K. Zurek and G. T. Gray III, in Shock Compression of Condensed Matter-1995, Eds. S.C. Schmidt, and W.C. Tao, AIP Conference Proceedings 370, New-York, 1996, p.643.

[7] S.I. Hong, G. T. Gray III, and J.J. Lewandowski, Scripta Met., 27, 431, (1992).

[8] S.I. Hong, G. T. Gray III, and J.J. Lewandowski, Acta Met., 41, 2337, (1993).

[9] M. Stuivinga, and E.P. Carton, in Shock Compression of Condensed Matter-1997, Eds. S.C. Schmidt, D.P. Dandekar and J.W. Forbes, AIP Conference Proceedings 429, New-York, 1998, p.619.

[10] D. Muscat, K. Shanker, and R.A.L. Drew, Material Science and Technology, 8, 971, (1992)

[11] W.R. Blumenthal, G. T. Gray III, and T.N. Claytor, Journ. Mater. Sci., 29, 4567, (1994).

[12] B. Klein, N. Frage, M.P. Dariel and E. Zaretsky, in Shock Compression of Condensed Matter-2001, Eds. M. D. Furnish, N. . Thadhani and Y. Horie, AIP Conference Proceedings 620, New-York, 2001, p.1119.

[13] B. Klein, N. Frage, M.P. Dariel and E. Zaretsky, J.Appl.Phys., 93, 968, (2003).

[14] L.M. Barker, and R.E. Hollenbach, J.Appl.Phys., 43, 4669, (1972).

[15] Ya.B. Zel'dovich, and Yu.P. Raizer, Physics of Shock Waves and High-Temperature Hydrodynamic Phenomena, Vol. 1, Academic Press, New-York and London, 1966.

[16] The slope of the Hugoniot hydrostat of Armco-iron is $S = 1.33$ [L.M. Barker, and R.E. Hollenbach, J.Appl.Phys., **45**, 4872, (1974)]. Since no data on S of TiC were available, we used for estimating S an expression [J.D. Johnson, in Shock Compression of Condensed Matter-1997, Eds. S.C. Schmidt, D.P. Dandekar and J.W. Forbes, AIP Conference Proceedings 429 (New-York, 1998) p.27] that relates the initial Hugoniot slope with the density derivative of the bulk modulus B_s, $S = 1 + (\rho_0 / B_s)(\partial B_s / \partial \rho_0)_s$. The derivative was calculated using the semi-empirical dependence of the modulus on the density $B_s = 2.77 \cdot 10^6 \rho^{4/3}$ [O.L. Anderson, in "Physical Acoustics" Ed. W.P. Mason, (Academic Press, New York – London, 1965) Vol. IIIB, p. 43]. The calculations yield for the initial Hugoniot slope of TiC the value $S = 1.35$. Thus, one single value, $S = 1.35$, was adopted for all the materials that were looked at.

[17] Z. Rosenberg, J. Appl. Phys., **74**, 752 (1993).

[18] Z. Rosenberg, J. Appl. Phys., **76,** 1543 (1994).

[19] R.D. Cook and W.C. Young, Advanced Mechanics of Materials, Macmillan Publishing Company, NY, 1985

[20] J. Aboudi, Int. J. Eng. Sci., **25**, 1229 (1987).

[21] J. Aboudi, Applied Mech. Rev. **49**, S83 (1996).

[22] H.C. Heard and C.F. Cline, J. Mat. Sci., **15** 1889 (1980).

[23] M.F. Ashby and C. G. Sammis, Pure and Appl. Geophysics , **133**, 489 (1990).

Ceramic Armor and Armor Systems

Ballistic Testing Study and Ballistic Performance of Ceramic Armor and Armor Systems

FAILURE OF PROJECTILE IMPACT RESISTANT GLASS PANELS

Richard C. Bradt, Stanley E. Jones, Mark E. Barkey and Michael E. Stevenson*
College of Engineering
The University of Alabama
Tuscaloosa, AL 35487 - 0202 USA

ABSTRACT

Projectile resistant glass panels are considered from the perspective of their glass/polymer laminate structure and processing. Design of these laminated glass panels is addressed from the review of prior and current studies. This establishes the basis for addressing the cracking patterns that develop under severe impact conditions. The sequence of failure mechanisms extending from those of modest impacts by granite rock projectiles, similar to automobile windshield stone impacts, to those of 30 caliber rifle ballistic impacts are discussed. Favorable design characteristics of the glass/polymer laminate front impact side glass and the central polymer layer for enhanced impact resistance are discussed. Suggestions are proposed for improved impact damage resistance design.

INTRODUCTION

The utilization of impact resistant, also known as projectile, or bullet resistant glass plate, or panel is increasing for reasons that are evident. Most individuals would be surprised to know just how much impact resistant glass paneling they regularly encounter. However, usually they are not aware of its presence, much less of its structural character. Almost daily, one is in the proximity of this type of glass. In one form or another, this glass product is present in car and aircraft windshields and in side panels of bus stops in many cities. It is in the display cases in stores, many doors and panels where security is an issue as in banks and retention facilities, and, of course in the windows of armored vehicles. It is not a new product and it is readily commercially available as can be ascertained from the many internet listings. However, there is little technical information in the open literature about it.

* Currently with MME, 1039 Industrial Court, Suwanee, GA 30024.

It is well known that glass will break easily in many applications. However, glass is also a material which can be applied effectively in designs to resist impact. It can inhibit or defeat the transmission of projectiles while maintaining a high level of transparency. Of course, transparency is only relative, as glass panels that are specifically designed for high energy impact resistance can be quite thick and the presence of minor impurities in the glass may have considerable absorption in the visible spectrum. These panels can be detected by the trained technician, although their presence will usually remain unknown to the casual observer.

STRUCTURE OF PROJECTILE / IMPACT RESISTANT GLASS PANELS

Figure 1 illustrates the general structure of an impact resistant glass panel. In its simplest form, it may be described as a laminated glass / polymer structure[1]. It is the thickness and the numbers of layers in the laminate that determines the energy absorption, impact resistance, or the projectile stopping power of one of these glass panels. For high intensity impacts the numbers of layers and their thicknesses can be substantial, so the weight per unit area is significant. It must be noted that the extraordinary impact properties of laminates are not restricted to only glass / polymer laminates. Extensive research has been completed on alumina / aluminum[2] and graphite / epoxy[3], as well as aluminum[4] and polymer[5] laminates. During some of these studies, it has been recognized that there are size effects present in the structure of these types of unique impact resistant composites[6].

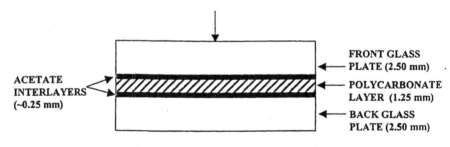

IMPACT FRONT SIDE OF LAMINATED GLASS PANEL

ACETATE INTERLAYERS (~0.25 mm)

FRONT GLASS PLATE (2.50 mm)
POLYCARBONATE LAYER (1.25 mm)
BACK GLASS PLATE (2.50 mm)

BACK SIDE OF LAMINATED GLASS PANEL

Figure 1. Crosssection view of a typical impact resistant glass laminate.

The Underwriters Laboratories have published a description of the impact resistance of different levels, or grades of impact resistant glass panel laminates[7]. However, it is not in the usual scientific terminology, such as might be based on its energy absorption. It is in terms of the categories of firearms it will defeat, such as 22 caliber, 30-06, 357 magnum, etc. The UL description also has categories that refer to multiple impacts from the specific weapon that may occur within a specified surface area of the target panel. Unfortunately, this less than scientific approach has been continued in other descriptions of impact resistant glass panels such as the Florida 2" x 4" beam code for hurricane resistant glass[8].

From the simple sketch of the glass panel laminate in Figure 1, there are several important features of these composites that merit technical discussion. This particular panel is a three layer symmetric laminate of glass / polycarbonate / glass, with a thin interlayer of acetate bonding. Other polymers may constitute the central layer in other laminates. It is of the type that might be used in an application where the hardness and other properties of the glass are required externally on both sides of the panel, yet significant impact resistance is required as well. As noted by McMaster, et al.[1], this form of laminate also eliminates extensive shard generation during impact for the glass remains bonded to the polymer. The cascadation type of fracture from a projectile impact of just such a panel, one that was used on a big city bus stop has been described at length in the literature[9]. The three major layers, the front glass, the central polymer layer and the back glass are bonded by a separate polymer interlayer, often an acetate. During manufacture, the layers of the composite are bonded either in a press type of device at a moderately elevated temperature or assembled in an autoclave. It is essential that there are no voids, or delaminations between the various composite layers. It is also critical for the optics that there is no dirt or other form of contamination between the layers. This latter requirement may demand manufacturing in a clean room for the highest quality transparent panels.

As addressed when considering the laminate failure, the front impact-side glass plate, or layer, and the backside glass plate exhibit different mechanical responses and failure modes on impact. Consequently there is some logic to consider having the front and the back surface layers possess different properties. Each will be briefly discussed. The front side receives the impact and normally fails through a fracture sequence that begins with a Hertzian-like crushing of the glass beneath the impacting projectile. This is accompanied by the flexure of the panel or front glass plate. Thus there is considerable benefit to have the impacting side consist of a much higher hardness material than common soda-lime-silica glass ($\sim$600 kgm/mm^2). This may be accomplished with the use of transparent ceramics such as single crystal sapphire ($\sim$2,000 kgm/mm^2), a dense polycrystalline alumina, an AlON, or other high hardness, high elastic modulus technical ceramics. Incorporation of such a layer into the laminate composite structure may create additional processing challenges, but the benefits are obvious at the point of damage initiation.

The backside glass plate of the laminate normally experiences considerable spalling and expels fine glass debris away from the backside during high velocity impact failures. This backside, or behind armor debris, can be highly detrimental to individuals or objects behind the panel and must be avoided. It has received considerable study for other armor systems[10-12]. Back side debris can be minimized in these glass panels by having the back side layer of the laminated panel to be a layer of polycarbonate, or another suitable polymer. It will strongly bond to the glass fragments generated during a total perforation event and will minimize, or perhaps even totally eliminate the back side debris. This involves a design tradeoff or compromise, as the much softer polycarbonate, when on an external layer is susceptible to scratching and other forms of surface degradation. These surface disturbances may create diffuse reflections and decrease the transparency.

While the two above design considerations are obvious ones, they cannot always be accommodated. The use of a dense transparent high hardness ceramic for the front layer is expensive compared to common soda-lime-silica window glass. The applications of backside layers of polycarbonate to reduce, or eliminate behind armor debris cannot always tolerate frequent contacts with the back side surface because of its softness. When the environment does not allow this transparency degradation, then a compromise must be achieved.

IMPACT TESTING AND THE FRACTURE OF GLASS PANEL LAMINATES

The experimental testing of glass laminates for projectile resistance has not been extensively reported in the literature. There are, however, several systematic studies that have provided considerable insight to the failure process[13-17]. These have contributed information valuable to the design methodology of glass panel laminates as well. One is the low velocity impact of laminated panels to simulate a stone impact of automobile windshields, or to simulate high winds blowing stones against architectural glass. The second is a higher velocity ballistic impact study of three layer laminates by a 30 caliber projectile. Both types of studies report testing of increasingly severe impacts that enable the researcher to follow the evolution of the glass damage. Each reveals different, but complementary aspects of the panel failure processes. They are discussed from their individual perspectives and overlap, incorporating the results of other peripheral studies that complement them as well.

Prior to those discussions, it is desirable to briefly address the experimental aspects of the impact testing of glass plates, or laminated glass/polymer composite panels. This has implications for the complete glazing system design of impact resistant transparent glass panel installations. The first item in the experimental testing is to choose the test panel size. It has been noted that there are size effects on the properties of composite laminates. From the mechanics perspective, larger panels of similar structure and thickness may be expected to experience greater flexure during impact. It is also necessary to hold or restrain the glass laminate test panel during the actual test, which is the method of glazing in the glass window industry. It is critically important to have a mounting device such that the test specimen is not damaged from contact, or clamping at the edges. Some form of rubber support system is appropriate about the edges. These features of the experiments cannot be overemphasized as Wilson[18] has shown that for low velocity impacts the clamping of monolithic glass impact specimens directly influences the resulting fracture, or cracking patterns of the plate. It is a simple extension that the same should apply to composite glass laminates in principle, although perhaps not in direct translation.

One must make a choice of projectile parameters as well. The critical point is that these many impact related parameters are not independent and may be expected to affect the experimental results. For these reasons, there is an ever increasing interest in the computer modeling and simulation of impact events[15]. However, even with these capabilities, there is considerable merit to actual full size application testing of the final design.

While the many computer simulations of the glass panel impact problem are inappropriate to review here as the topic is the glass failure or fracture, several are noteworthy of mention for readers who may wish to pursue the topic. In the low impact velocity regime, Flocker and Dharani[19,20] have applied finite element analyses to an architectural glass laminate. Not surprisingly, they determined that the maximum principal tensile stress occurred in the back side ply, on a line directly behind the impact point, the projectile trajectory. It is not surprising that radial cracking and failures have a tendency to initiate at that location for the impact side and the backside glass plates of the laminate. Of course, for any simulation of glass fracture, one must have a failure criterion for the glass. In the context of computer simulations, this issue has been addressed by Glenn[21] and by Bouzid, et al[22]. Glenn advanced the generally accepted concept of totally brittle failure of the glass originating from the point of maximun principal tensile stress, which appears to agree with the initiation of the radial cracks from the flexure of the impact event.

Low Velocity Impacts

The Grant, et al.[13] study of the initiation of damage of automobile windshields by low velocity granite stones is an excellent example for addressing damage initiation to the impact side glass layer. During mild or low velocity impacts, it is often possible for the outer glass plate, or layer on the impact side to incur severe damage and cracking while subsequent glass layers of a multilayer laminate are not damaged at all. This illustrates the merits of the low modulus polymer layer separation, which allows the flexure of the entire panel, often without any damage to the back side glass layers. Behr, et al.[16] have addressed the role of an internal PVB layer and concluded, not surprisingly, that the impact resistance of the laminate glass panel was increased when the polymer layer thickness was increased. The noted scenario of an undamaged back side glass in a three layer laminate is almost always the case for impacts near to the velocity, energy or momentum conditions for the initiation of damage the threshold condition for the impact side glass of one of these composite laminates.

Grant, et al.[13] have completed an extensive study for a three layer laminate with outer layers of standard float glass separated by a central layer of PVB, polyvinyl butyral. It is perhaps the classic study for an understanding of the role of the front impact glass layer of a laminate. They varied the front glass layer thickness by nearly a factor of four, while keeping the PVB central layer thickness constant. The study confirmed two important effects: (i) an increase in the threshold projectile velocity to initiate damage with increasing thickness of the front glass layer, and (ii) a transition in the damage mechanisms, or damage character of the front glass layer with increasing thickness of that glass layer. The first of these is not surprising, for one naturally expects greater resistance from thicker layers, similar to the previously noted observation for a thicker central polymer layer. However, increased thickness also translates to increased weight and increased cost, so that improvement may not always be considered to be totally beneficial from all perspectives.

The Grant, et al.[13] observation of changes of the glass damage mechanism as the front layer thickness increases demonstrates the value of systematic study. Different types of damage patterns for the front glass layer were identified as the impact velocity was increased. At the lowest impact velocities, the front glass plate first experienced mild crushing beneath the contact area of the impacting stone projectile. For the thin front glass plates, after crushing, the flexure of the glass laminate created the classical star-like crack pattern associated with impacts. This star initiated from the internal surface of the outer glass plate. It was created during the flexure of that outside glass plate during the impact. Frequently at higher impact velocities, Hertzian cone cracks formed. Star cracking and cone cracking were frequently observed together for slightly higher velocity impacts. Although the maximum stone velocities in the Grant, et al.[13] study were only ~30 m/s, hardly a high velocity impact, their results are nonetheless an outstanding contribution to an understanding of the initiation of the failure process of impacted glass laminates.

Pantelides, et al.[17] studied the the same velocity range of impacting stones, a study that was spawned in the aftermath of hurricane Alicia thrashing Houston in the early 1980's. The 100 mile per hour winds blew the chat-like stones from the roofs of adjacent buildings and broke the monolithic glazing from the walls of skyscrapers. They studied a three layer laminate with heat strengthened glass and a central polymer layer of PVB. Then they also studied a three layer polymer composite with heat strengthened outer layers of glass. For the latter, a layer of PET, polyethylene terepthalate, was sandwiched between two PVB layers. There were two unique features to this study, the heat strengthening of the glass and the multiple polymer layers. From a materials design perspective, it would appear that both of these features would be beneficial to the general impact resistance of the panels. The heat strengthened glass should inhibit damage initiation and the multiple composite layers should enhance the total energy absorption, once severe damage was the result.

Heat strengthening creates a modest compressive stress at the surface of the glass plates. A surface residual compressive stress might be expected to reduce the damage from impacting projectiles on the impact face of the outer glass plate. This would also increase the amount of flexure that the glass plate is able to experience without the initiation of the star fracture pattern from the accompanying flexure of the impact event. However, although strengthening by the incorporation of residual stresses may inhibit damage initiation, whenever residual stresses are present, they will also increase the stored elastic strain energy of the object. As the elastic strain energy is proportional to the stress squared, if fracture is initiated, then it will have considerably more energy to feed the cracking processes associated with the failure event. Strengthening can thus be a two edged sword!

As for the internal three layer polymer laminate, it was previously mentioned that polymer laminates have been studied for armor applications. Results of the Mines, et al.[5] study suggest that the incorporation of a multilayer polymer laminate between the two outer glass plates should remarkably improve the impact resistance. Indeed it did, as the most damage resistant of the panels that Pantelides, et al.[17] tested were those with the tri-layer, central, PVB/PET/PVB sandwich between two outer glass plates. In retrospect, this is not a surprising observation.

Prior to considering the higher velocity projectile impacts, it is important to summarize these results for low velocity impacts with a variety of projectile materials from granite stones to steel balls. It appears that the damage initiation to the sandwich panel laminate begins with the crushing of the outer glass layer directly beneath the impacting projectile. This may institute a Hertzian-like cracking pattern beneath the projectile contact point. It then evolves into the classical impact star cracking pattern by the production of the radial cracks from the inner surface of the outside glass plate as a result of the flexure of that plate with the entire panel flexure during the impact event. These events will usually proceed without any damage whatsoever to the backside glass plate.

There appear to be options to improve the impact resistance of these impact resistant glass/polymer laminate structures. These include increasing the thickness or strength of the impact side glass plate, increasing the thickness of the central polymer layer, or enhancing the single central polymer layer to a multilayer polymer composite section within the overall panel. Each of these has compelling evidence for improved impact damage resistance.

High Velocity Impacts

The study of Stevenson, et al.[14] for 30 caliber projectiles impacting a simple glass laminate identical to the one depicted in Figure 1 is the most extensive study of this type of glass panel ballistic impacts in the public domain. They addressed the development of the fracture patterns, the cracking of the front and back glass layers of a three layer laminate for a range of projectile velocities and kinetic energies. These were well beyond those examined for the low velocity rock and stone impacts just described. However, the progressive nature of the glass damage and fracture pattern development is remarkably similar for both levels of impacting events.

Figure 2 illustrates the evolution of the cracking pattern on the front glass plate and Figure 3 that of the backside glass. The slowest impact velocity was 48 m/s for the ~11g copper projectile. Even though the velocities overlap with the previous studies, the kinetic energy of the projectile was significantly higher even for the lowest velocity. Even at this low velocity, all three of the layers experienced some damage. The front glass plate showed evidence of Hertzian crushing of the glass beneath the projectile at the point of impact. This damage was similar to that previously reported by Kirchner and Gruver[23] for monolithic glass impacts.

At this velocity, the central polycarbonate layer experienced some delamination and the back glass plate contained a few short radial cracks. This suggests a difference in the origins of the radial cracks in the two plates. For the front glass plate, the development of radial cracking was preceded by the Hertzian-like crushing of the glass by the projectile. Radial cracking on the backside plate occurred without any crushing or direct projectile impact contact with the glass plate. This may be a situation of greater damage to the tensile bending side of the panel during impact. No fragmentation, dicing or spalling of the glass occurred on either side of the panel in the area surrounding the impact site.

With an ~ 50% increase in the projectile velocity to 66 m/s, the damage to the glass panel increased substantially, but the panel did not experience perforation. Radial cracking increased on both the front and back glasses, and considerable glass spalling occurred in the glass panels on both sides. Numerous circumferential cracks were evident. Interestingly, the backside glass spalling was of a larger diameter than that of the front glass plate. This may suggest that the spalling is related to the flexural displacement as the backside achieves maximum flexure during the impact, but the front side glass plate probably achives its maximum flexure during the springback of the panel. There was significant delamination of the central polymer layer from the outer glass plates and a limited punching through of the polymer by the projectile. This further confirms that as the damage increases, the damage processes for the front and back glass plates are different.

At 74 m/s, only about 10% faster than the previous impact, spalling was substantial for both glass plates, front and back, but the spalled area appeared to decrease slightly from that for the 66 m/s impact. A plug of the polymer layer was pushed through the panel and protruded from the backside, but the laminate panel did not experience complete perforation by the copper projectile. The toughening aspects of the central polymer layer, a polycarbonate one in this instance, is readily apparent.

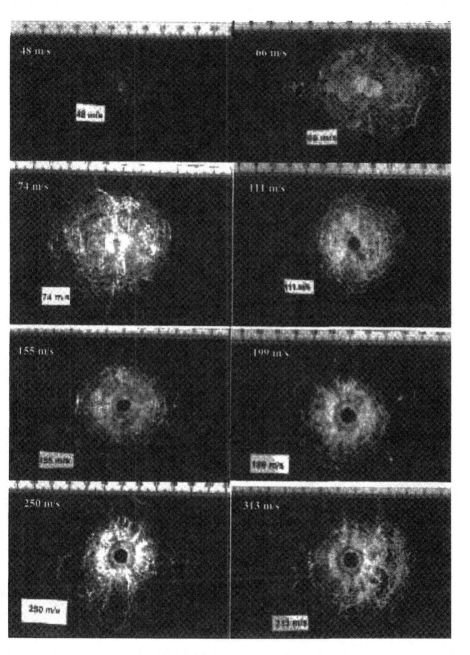

Figure 2. The front impact sides of the projectile impacted glass panel laminates. The velocities are in the white squares on the left and range from 48 m/s to 313 m/s. Note that the damaged area seems to achieve a maximum size at about 74 m/s, then decreases in size with further projectile velocity increases.

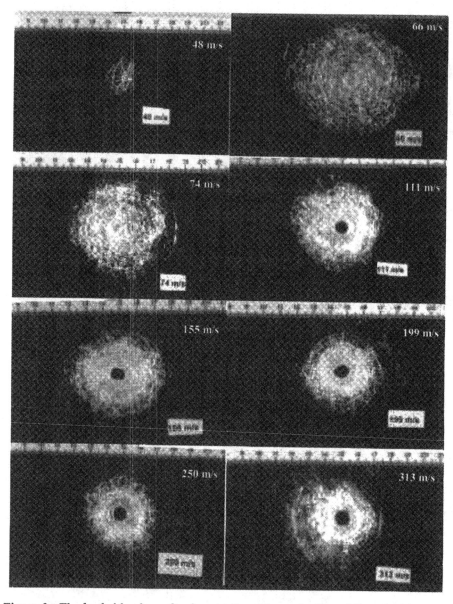

Figure 3. The backside glass after impact from the projectiles. The velocities are show in the white squares on the right and are the same as for the front glass impact patterns in Figure 2. Compare with the front side cracking pattern in Figure 2. Note that both appear to experience a maximum in their localized damage area at an intermediate projectile velocity.

Ceramic Armor and Armor Systems

For the next higher projectile velocity, 111 m/s, the projectile completely penetrated the laminate panel. However, the size of the damage zone appeared to decrease even further from the maximum diameter. At 155 m/s the glass laminate panel was also completely penetrated by the projectile. The damage zone about the perforation continued to decrease in diameter. A new form of cracks appeared in the front side glass plate at this projectile velocity. It was the appearance of wing-like shaped cracks that were inclined at ~45° to the plane of the glass plate. These were just outside of the spall zone, but only in the front glass plate at first. Further increases in the projectile velocity produced more of these wing-like angled cracks and also smaller apparent damage zones of the spalling. One might imagine that at very high velocities, beyond the maximum achieved in this study, the projectile might just punch through the laminate panel with only a minimum of collateral damage.

Higher projectile impact velocities, >155 m/s, increased the numbers of the wing-like cracks and also the size of the actual perforation hole in the glass panel. This is not surprising for the projectile itself also experienced increased damage and mushrooming with increasing projectile velocities. The sizes of the spalling zones on both glass plates decreased through the 313 m/s maximum velocity achieved.

Sequence of Glass Panel Damage
One can approximately describe the sequence of damage events leading to the defeat of the glass panel by the projectile, although the precise details of the sequence are not completely clear. Upon impact of the panel, the front side glass plate experiences some crushing and Hertzian-like damage beneath the contact area of the projectile. The entire panel begins to flex and radial cracks develop from beneath the contact area on the inner surface of the front side glass plate. It is not completely clear if these radial cracks intiate before the Hertzian crushing reaches the inside glass plate surface, or if they are only from the flexure of the front glass plate. It is not obvious how the local and global fracture modes are separated. In any event, at this point the front glass plate in the three layer laminate is experiencing substantial cracking and damage without the backside glass plate experiencing any noticeable degradation. The damage seems evolve in this fashion for both the lower velocity and the higher velocity ballistic impacts. In other words, constructing more robust panels for ballistic impacts does not appear to alter the sequence of damage evolution.

The next events are the delamination of the polymer from the glass plates and the initiation of severe circumferential cracking. The projectile punches through the polymer layer and initiates damage on the backside glass plate. Major spalling is evident for both the front and back side glass plates. The panel experiences perforation by the projectile. Following complete penetration, or perforation of the panel, the spalling zones begin to decrease in size as the projectile velocities increase. The 45° wing-like cracks appear, first on the front glass, but only for higher velocities on the backside glass. As the velocities increase, the actual penetration hole in the panel also increases. However, the flattened mushroom impact end of the projectile is always larger in diameter than the resulting hole as discussed by Stevenson, et al.[14] This indicates that there is significant elastic recovery of the laminated panel from its state of extended flexure when the projectile actually penetrates the structure.

It is clear that the similar damage events occur slightly differently and often later during the impact event in the backside glass plate than on the front side glass plate. The two most obvious reasons for this conclusion are the points that the front glass plate experiences considerable damage of a Hertzian-like crushing beneath the projectile and also the initiation of radial cracking prior to any damage to the backside glass plate. The second is the general velocity of appearance for the 45° wing-like cracks. In the front glass plate, these unique cracks are first seen at about 155 m/s and continue to increase in density as the velocity increases. In the backside glass plate, these particular cracks do not appear in evidence until the 313 m/s impacts.

One must conclude that the damage and failure mechanisms are definitely different for the front and back glass plates in a three layer glass/polymer laminate. This suggests that it may be advantageous to apply slightly different design criteria to the two different plates for enhanced impact resistance. As noted by UL[7], the impact resistant glass laminates that resist the more severe projectile impacts will consist of multiple layers of glass and polymers, perhaps as many as a half dozen of each. Of course the thicknesses of these layers may vary, too. It is evident that the crack pattern development, or the failure criteria for the successive glass layers in a multilayer panel may reasonably be expected to be different. The extent of this difference has not been experimentally, nor theoretically investigated to date. Neither have its benefits been incorporated into design modifications. However, the incorporation of design differences to accomodate, or combat the failure differences of internal plates has many intriguing design options that have not been applied.

SUMMARY AND CONCLUSIONS

The sequence of damage to polymer/glass panel laminates for impact resistance to projectiles was addressed for one of the simplest of these composites, a glass/polymer/glass three layer system. The range of projectiles that have been considered extend from the low velocities of stones strinking automobile windshields and architectural glass impacted by airborne stones during windstorms to the higher velocities of a copper projectile from a 30 caliber rifle. All were considered within the context of examining increasingly severe impact situations, thus creating some overlap to the ranges of impact severity. Several characteristics of the glass and the central polymer layer that contribute to increased impact damage resistance were identified and have been briefly discussed. Speculations for the extension of several of the design concepts to multiple layer glass/polymer composite panels are proposed.

ACKNOWLEDGEMENT

The authors thank Mr. Hidemi Nakai of the Nippon Sheet Glass Company, Tokyo, Japan for graciously supplying the laminated glass panels for testing. Technical discussions and assistance with the testing are gratefully acknowledged with W. Rule, J. Wagner, D. Ahern and K. Black.

REFERENCES

[1] R.A. McMaster, D.M. Shetterly and A.G. Bueno, "Annealed and Tempered Glass", 453-459 in *Ceramics and Glasses,* Engineered Materials Handbook, Vol. IV, ASM-I, Materials Park, Ohio (1991).

[2] B.A. Roeder and C.T. Sun, "Dynamic Penetration of Alumina / Aluminum Laminates: Experiments and Modeling", *Int. J. of Impact Eng.* 25, 169 - 185 (2001).

[3] J.A. Nemes, H. Eskandari and L. Rakitch, "Effect of Laminate Parameters on Penetration of Graphite / Epoxy Composites", *Int. J. of Impact Eng.* 21, 97 - 112 (2001).

[4] R.L. Woodward and S.J. Cimpoeru, "A Study of the Perforation of Aluminum Laminate Targets", *Int. J. of Impact Eng.* 21, 117 - 131 (1998).

[5] R.A.W. Mines, A. M. Roach and N. Jones, "High Velocity Perforation Behaviour of Polymer Composite Laminates", *Int. J. of Impact Eng.* 22, 561 - 588 (1999).

[6] D. Liu, B.B. Raju and X. Dang, "Size Effects on Impact Response of Composite Laminates", *Int. J. of Impact Eng.* 21, 837 - 854 (1998).

[7] *Bullet Resistant Equipment,* Underwriters Laboratory Report UL 752, 9th Edition, Underwriters Laboratories, Inc., Northbrook, IL (1995).

[8] *South Florida Building Code*, 1 - 9, Impact Test Procedures, Protocol PA-94, Dade County Building Code Compliance Office, Miami, FL (1994).

[9] T. Sakai, M. Ramulu, A. Ghosh and R.C. Bradt, "Cascadation Fracture in a Laminated Safety Glass Panel", *Int. J. of Fracture* 48, 49 - 69 (1991).

[10] F. Horz, M.J. Cintala, R.P. Bernhard and R.H. See, "Evolution of Debris Plumes as Inferred from Witness Plates", *Int. J. of Impact Eng.* 20, 387 - 398 (1997).

[11] A.L. Yarin, I.V. Roisman, K. Weber, and V. Hohler, "Model for Ballistic Fragmentation and Behind-Armor Debris", *Int. J. of Impact Eng.* 24, 171- 201 (2000).

[12] W. Arnold and W. Paul, "Behind Armor Debris investigation and their application into a New Vulnerability Model", *Int. J. of Impact Eng.* 26, 21 - 32 (2001).

[13] P.V. Grant, W.J. Cantwell, H. McKenzie and P. Corkhill, "The Damage Threshold of Laminated Glass Structures", *Int. J. of Impact Eng.* 21, 737 - 746 (1998).

[14] M.E. Stevenson, S.E. Jones and R.C. Bradt, "Fracture Patterns of a Composite Safety Glass Panel during High Velocity Projectile Impacts", 473 - 488 in *Fractography of Glass and Ceramics IV,* edited by J.R. Varner and G.D. Quinn, Ceramic Trans. 122, The American Ceramic Society, Westerville, OH (2000).

[15] R. C. Bradt, M.E. Barkey, M.E. Stevenson and S.E. Jones, "Projectile Impact-A Major Cause for Fracture of Flat Glass", *The Glass Researcher, Bulletin of Glass Science and Engineering* 11, 20 - 23 (2002).

[16] R.A. Behr, J.E. Minor and H.S. Norville, "Structural Behavior of Architectural Laminated Glass", *J. Struct. Eng.* 119, 202 - 222 (1993).

[17] C. P. Pantelides, A. D. Host and J.E. Minor, "Postbreakup Behavior of Heat Strengthened Laminated Glass under Wind Effects", *J. Struct. Eng.* 119, 454 - 467 (1993).

[18] J.F. Wilson, "Similitude Experiments on Projectile-Induced Fracture of Monolithic Glass", *Int. J. of Impact Eng.* 18, 417 - 424 (1996).

[19] F.W. Flocker and L.R. Dharani, "Stresses in Laminated Glass subject to Low Velocity Impact", *Eng. Structures* 19, 851 - 856 (1997).

[20] F.W. Flocker and L.R. Dharani, "Low Velocity Impact Resistance of Laminated Architectural Glass", *J. Architect. Eng.* 4, 12 - 17 (1998).

[21] L.A. Glenn, "The Fracture of a Glass Half-space by Projectile Impact", *Phys. Solids* 24, 93 - 106 (1976).

[22] S. Bouzid, A Nyoungue, Z. Azari, N. Bouaouadja and G.Pluvinage, "Fracture Criterion for Glass under Impact Loading", *Int. J. Impact Eng.* 25, 831 - 845 (2001).

[23] H.P. Kirchner and R.M. Gruver, "Localized Impact Damage in a Viscous Medium", *Fracture Mechanics of Ceramics* 3, 365 - 377, edited by R.C. Bradt, et al., Plenum Pub. Co., NY, NY (1977).

PENETRATION ANALYSIS OF CERAMIC ARMOR BACKED BY COMPOSITE MATERIALS

Moshe Ravid
Rimat Advanced Technologies, Ltd.
8B Simtat Hayerek St.
Hod Hasharon 45264, Israel.
e-mail: rimat2@netvision.net.il

Sol R. Bodner
Technion - Israel Institute of Technology
Haifa 32000, Israel.
e-mail: mersbod@techunix.technion.ac.il

Sidney Chocron
Engineering Dynamics Dept.
Southwest Research Institute
San Antonio, Texas 78228-0510
USA.
e-mail: Sidney.Chocron@swri.org

ABSTRACT

This paper is concerned with the further development of the multi-stage penetration mechanics model of ceramic armor originally proposed in 1989 by Ravid & Bodner Ref. [1] and revised in 1999 Ref [2]. The 2D analysis relies on the treatment of the initial shock stage (1987) Ref. [3], and on the model of 1983 for a rigid projectile penetrating a viscoplastic target Ref. [4]. After the shock stage that leads to shattering of the ceramic layer, continued penetration of the projectile into fragmented ceramic, held in place by the backing plate and by inertial effects, is an important part of the process. The final stage involves deformation and penetration of the backup plate which can consist of Aramid or high density Polyethylene in polymeric matrices. The performance of such laminate backing materials under impact loading has been studied which included detailed examination of their straining and failure. These physical considerations are incorporated into the final penetration stage of the current analysis of ceramic armor.

INTRODUCTION

During the past decade, considerable attention has been given to the mechanical behavior of ceramics under impact and thermal loadings. These investigations have encompassed the full range of dimensional scales, and some of the results have been applied to ceramic armor. In the simplest case, this armor consists of a frontal layer of ceramic tiles backed by plates of a ductile metal or of composite materials. Present practice for the support plate tends to the use of Aramid or high density Polyethylene in polymeric matrices. Examination of the behavior of such backing plates is therefore a necessary part of the investigation.

The acquired information is intended to be used in computer simulations of the ballistic event and also in the development of analytical models of the process. Both approaches encounter difficulties due to uncertainties in understanding essential physical mechanisms such as the conditions and process of dynamic failure of ceramics, the ballistic resistance of fragmented ceramic held in place by a support plate, and the performance of the backing plate as an integral component of the armor system.

The present paper is a continuation of the development of the analytical models of [1] and [2] for the ballistic penetration of ceramic armor. Emphasis here is given to a detailed examination of the performance of laminated composite backup plates that are currently in use.

ANALYTICAL MODEL

In [1] and [2], the initial stage of impact is characterized by shock waves developed in the ceramic frontal layer and in the projectile as described in [3]. When the shock wave in the ceramic reaches the interface with the backup plate, rarefaction waves are generated which propagate backwards towards the impact surface. The shock compressed ceramic is more dense than the original state but had experienced damage in the form of microcracks developed by the initial wave. These effects influence the velocity of the rarefaction wave in opposing manners. After a short delay, the rarefaction wave is followed by breakup of the ceramic whose front can be considered to be a "shatter wave". The velocity of the shatter wave is estimated to be $c_0/2$ where c_0 is the velocity of bulk elastic waves and is approximately equivalent to that of the Rayleigh surface wave for maximum crack propagation.

The shatter wave velocity should also be influenced by the bonding at the interface and by the impedance mismatch effect on the intensity of the reflected wave. A factor F on the velocity $c_0/2$ is taken to be,

$$F = \frac{P_{H1} - P_{H2}}{P_{H1}} \tag{1}$$

where P_{H1} is the Hugoniot pressure in the ceramic due to projectile impact, and P_{H2} is the Hugoniot pressure delivered to the backup plate due to "impact" of the ceramic on it (at twice the initial particle velocity in the ceramic). Typical values for F are in the range $0.65 - 0.90$.

Complete breakup of the ceramic is presumed to occur when the front of the shatter wave reaches the forward extent of the "inelastic zone" of comminuted (pulverized) ceramic surrounding the imbedded projectile. Penetration of the projectile during the shock stage is calculated from [3] up to a time T_1 when shock effects are diminished. After T_1, penetration continues according to the analysis of [4] and projectile erosion, as determined in [3], is assumed to continue at the same rate until breakup. Subsequent to breakup, continued penetration is resisted by fragmented ceramic pieces held in place and constrained by the backup plate and by inertial effects. The effective strength of the fragmented ceramic σ'_{0c} is estimated by an empirical equation,

$$\sigma'_{0c} = 0.2\,\sigma_{0c}\left\{1 - \exp\left[\frac{-m\sigma_{0b}}{0.2\,\sigma_{0c}}\left(\frac{H_b}{H_c}\right)^2\right]\right\} \tag{2}$$

In eq. (2), σ_{0b} is the in-plane strength of the backup layer of thickness H_b, and H_c is the ceramic thickness between the current projectile front and the interface. The modifying term in (2) depends on the relative inelastic bending moduli of the two components of the armor. For strong support, the effective strength would be 0.2 of the compressive strength of intact ceramic σ_{0c}, which was indicated in a few exploratory tests. Alternatively, weak backing of the fragmented ceramic would lead to low effective strength. The quantity m is intended to indicate the quality of bonding and is taken to be unity for optimum bonding.

Penetration of the projectile through fragmented ceramic is considered to be operative until the front of the inelastic zone surrounding the projectile reaches the interface. At this condition, motion of the backup plate is initiated with a velocity field corresponding to simple radial flow emanating from a point on the centerline as in the initial bulging mode of the model of [4]. For this stage, shear strength and frictional effects in the comminuted ceramic forward and moving with the projectile and in the laminated backing plate are ignored so that material displaced by the combined projectile bulges into the backing plate and leads to an equivalent volume bulge of that plate's outer surface. This mode is illustrated in Fig. 1 where the volume of material within the sector angle 2ψ for the spherical cap shaped bulge of the backup plate equals that for the bulge in the ceramic plate with sector angle 2β. As a consequence, the average velocity for each layer i of the backup plate with radial extent $\eta_i R$, $\overline{V}_i$, can be related to the current projectile velocity V and the radial flow geometry by,

$$\overline{V}_i = \frac{V}{3\eta_i^2} \frac{\left(1 - \cos^3 \psi\right)}{\left(1 - \cos \psi\right)}$$ (3)

where

$$\eta_i = \eta_b + \left(\eta_{bb} - \eta_b\right)\left[(i - 1)/N\right]$$ (4)

and the layers are numbered from i=1 to N, R is the projectile radius, and η_b, η_{bb} are shown in Fig. 1. According to the model of [5] for the behavior of fabric subjected to a transversely applied velocity, the average resisting force that each fabric layer, assumed elastic, would exert in the direction of the applied velocity would be

$$F_i = (E \, \varepsilon_i)\left[Y\left(\frac{\eta_i R}{\sin \psi}\right)\psi\right]S \cos\theta_i$$ (5)

where E is Young's modulus, ε_i is the strain, Y is the number of yarns of fabric per unit length, S is the cross section of the yarn, and θ_i is the angle between the force induced in the deformed yarn and the direction of projectile motion. It is also assumed that the force is constant over the effected length of each layer.

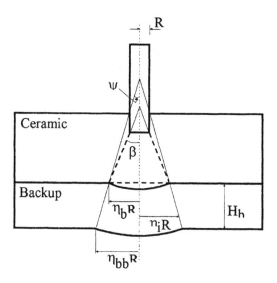

Fig. 1: Bulging of backup plate

An essential point in the model for the behavior of the laminated backup plate is that the resisting force in each layer, eq. (5), could be expressed in terms of the velocity of each layer, $\overline{V}_i$, which in turn is a function of the current projectile

Ceramic Armor and Armor Systems

velocity and the current geometry, eqs. (3),(4). The relevant equations are given in [6] and reproduced in [5], [7] and other papers. In the present notation they are,

$$\overline{V}_i = c\sqrt{\varepsilon_i\left(2\sqrt{\varepsilon_i(1+\varepsilon_i)} - \varepsilon_i\right)} \qquad (6)$$

where c is the longitudinal elastic wave velocity in the yarn. A practical value of c to be used for a strong fabric surrounded by a polymer matrix is.

$$c' = \left(c / \sqrt{2}\right)(1 - P)^{1/2} \qquad (7)$$

where the factor $\left(1/\sqrt{2}\right)$ is based on the work of [8] and P is the fraction of added mass of polymer. The $(1-P)^{1/2}$ term is approximately that obtained by calculating an effective modulus based on the law of mixtures. The angle θ_i is determined from [6] to be a function only of the strain ε_i so it could also be expressed in terms of V and the current geometry from (6) and (3),

$$\sin\theta_i = \frac{\sqrt{2\varepsilon_i\sqrt{\varepsilon_i(1+\varepsilon_i)} - \varepsilon_i^2}}{\sqrt{\varepsilon_i(1+\varepsilon_i)}} \qquad (8)$$

The total resisting force of the backup plate could be obtained by summing F_i for all the layers of fabric. However, the resisting force of each layer multiplied by its velocity $\overline{V}_i$ is summed and enters the overall work rate balance equation of [4] for determination of the current penetration velocity.

Continued penetration leads to a condition where the spherical cap geometry of the bulge of the ceramic layer changes to a bulge advancement mode as described in [4]. To simplify the calculation of ψ for subsequent penetration, the ratio (cosψ/cosβ) is held constant at the value existing at the transition. The velocity field described by (3), (4) remains unchanged. With further penetration, the projectile front reaches the original position of the rear surface of the ceramic layer. At this condition, the projectile and the comminuted ceramic material forward and moving with it are considered to act together as a rigid projectile having the current momentum. Penetration of the backup plate by the effective projectile would again be governed by the work rate balance equation of [4] using the preceding equations for the resisting force and velocity of each layer of fabric. During this final stage, the angles β and ψ are taken to be constant at the onset values. Failure of a layer of fabric is governed by a limiting strain criterion where the strains are obtained from eqs. (6) and (3). Until failure, each layer would contribute to the work rate balance equation.

EXAMPLES

Some numerical exercises were performed for the ceramic armor combinations described in [2]. The properties of the projectile, the ceramic outer plate, and those of the backup plate of Kevlar fibers in a polymer matrix are listed in Tables I and II. The examples consist of three different thicknesses of the ceramic layer, 8.5, 9 and 10 mm, and backed respectively by 28, 26 and 20 layers of Kevlar K770 laminate to provide for an equal areal density of 50 kg/m². Areal density of each Kevlar layer was 560 gram/m² with the matrix and 470 gram/ m² without the matrix.

Numerical exercises were carried out for each case for impact by a 7.62×54 mm API (Armour Piercing Incendiary) projectile at 870 m/s. Results of the penetration velocity-time history for each case are shown in Fig. 2. All three cases led to no perforation (similar to the results of the ballistic tests), but the one with the 10 mm ceramic (case 3 of Table II) indicated the best relative performance with no breakage of any of the layers of the backup plate.

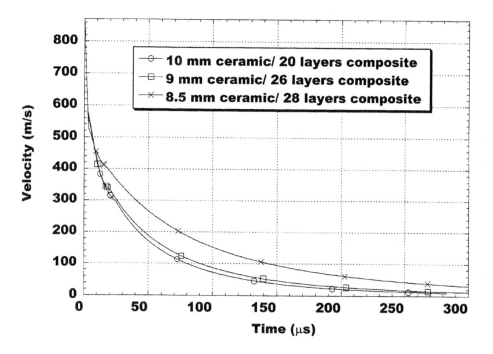

Fig. 2: Calculated penetration velocities for the three cases as functions of time.

Ceramic Armor and Armor Systems

TABLE I. MATERIAL PROPERTIES

Symbol	Units	Material Properties			
		Penetrator	$A\ell_2O_3$,98%	K 770 laminate	Kevlar Yarns
$\sigma_0(\dot{\varepsilon}_0 = 1s^{-1})$ flow stress	MPa	1765	2844	290 (average)	-
ρ density	kg/m^3	7850	3800	1400	1400
C $\dot{\varepsilon}$ coefficient	n.d.	0.025	0.04	0.02	-
ε_{max} failure strain	%	4	1	-	3.8
E Young's modulus	GPa	213	274	-	60
ν Poisson ratio	n.d.	0.31	0.22	0.45	-
c_0 bulk sound velocity	km/sec	4.55	10	(c) 2.07	(c) 6.55
α_1 shock wave constant	n.d.	1.45	1.3	-	-
Γ Gruneissen coefficient	n.d.	2.03	2.3	-	-

Cross section of yarn : $0.23\,\text{mm}^2$ ‖ yarns/m : 670 ‖ resin mass fraction : 0.125

$\sigma_y = \sigma_0[1 + C\log(\dot{\varepsilon}_{eff}/\dot{\varepsilon}_0)]$ ‖ Factor on shatter velocity : 0.70.

TABLE II. BALLISTIC TEST TARGETS AND CALCULATED RESULTS

Case	Layer 1		Layer 2		Number of K770 broken layers (calculated)
	Material	Thickness (mm)	Material	Thickness (mm)	
1	$A\ell_2O_3$ (98%)	8.5	K770	12.7 28 layers	12
2	$A\ell_2O_3$ (98%)	9	K770	11.4 26 layers	2
3	$A\ell_2O_3$ (98%)	10	K770	8.7 20 layers	0

projectile: 7.62×54mm R API-B32/0°N
total actual length of AP core: 29.9 mm
equivalent cylinder length: 23.5 mm

hard core diameter = 6.1 mm
hard core weight 5.39 gm
impact velocity: 870 ± 5 m/s
areal density of armor = 50 kg/m^2

DISCUSSION

The modification suggested in this paper of the analytical model for ballistic penetration of ceramic armor [2] leads to improved understanding of the response behavior of laminated backup plates. Those plates provide the support for fragmented ceramic to have effective resistance to ballistic penetration. They also serve to absorb kinetic energy of the residual effective projectile in the final stage of the penetration process. In that respect, the backup plate acts as a viscous medium, even for elastic fibers, since the resistance depends upon the imposed penetration velocity.

REFERENCES

[1]M. Ravid, S.R. Bodner and I. Holcman, "Application of Two Dimensional Analytical Models of Ballistic Penetration to Ceramic Armor", *Proceedings of the 11th International Symposium on Ballistics*, Brussels, Belgium, 1989.

[2]M. Ravid and S. Bodner, "Analytical Investigation of Ceramic Armor with Composite Laminate Backing", *Proceedings of the 18th Ballistic Symposium*, San Antonio, TX, USA, 1999.

[3]M. Ravid, S. Bodner and I. Holcman, "Analysis of Very High Speed Impact", *Int. J. of Engineering Science*, 25[4], 473-482 (1987).

[4]M. Ravid and S. Bodner, "Dynamic Perforation of Viscoplastic Plates by Rigid Projectiles", *Int. J. of Engineering Science*, 21[6], 577-591 (1983).

[5]I.S. Chocron-Benloulo, J. Rodríguez and V. Sánchez-Gálvez, "A Simple Analytical Model to Simulate Textile Fabric Ballistic Impact Behavior", *Textile Research Journal*, 67[7], 34-41 (1997).

[6]J.C. Smith, F.L. McCrackin and H.F. Schiefer, "Stress-Strain Relationships in Yarns Subjected to Rapid Impact Loading, Part V: Wave Propagation in Long Textile Yarns Impacted Transversely", *Textile Research Journal*, 228-302 (1958).

[7]I.S. Chocron-Benloulo and V. Sánchez-Gálvez, "A New Analytical Model to Simulate Impact Onto Ceramic/Composite Armors", *Int. J. of Impact Engineering*, 21[6], 461-471 (1998).

[8]D. Roylance, A. Wilde and C. Tocci, "Ballistic Impact of Textile Structures", *Textile Research Journal*, 34-41 (1973).

RESISTANCE OF DIFFERENT CERAMIC MATERIALS TO PENETRATION BY A TUNGSTEN CARBIDE CORED PROJECTILE

C. Roberson
Advanced Defence Materials Ltd.
Rugby,
Warwickshire.
CV21 3QP, UK.

P. J. Hazell
Cranfield University
Royal Military College of Science
Shrivenham, Oxfordshire
SN6 8LA, UK.

ABSTRACT

In this paper, the 7.62 × 51mm FFV* round consisting of a tungsten carbide core (Hv 1200) and copper gilding jacket was fired at silicon nitride, titanium diboride and boron carbide ceramics; the results were compared with associated work on silicon carbide. In each case, the ballistic performance of the ceramic was assessed by measuring the DoP into an a aluminium alloy witness block. The performance was then compared to micro-hardness values for each ceramic. Of particular interest is the measure of each ceramic's ability to damage and fracture the tungsten carbide core of the FFV round. This paper will be of interest to armour system designers and manufacturers.

INTRODUCTION

Tungsten carbide is well established as the penetrator material for the cores of special types of armour piercing ammunition. The former Soviet Union developed such ammunition for all their machine gun calibres over 40 years ago. Foremost amongst them the 14.5mm BS41 API round, which shaped the armour design of a generation of Infantry Fighting Vehicles fielded by NATO armies during the Cold War.

---

* FFV Ordnance now Bofors Carl Gustav AB.

Developments, starting in Sweden about 20 years ago, have now resulted in a subsequent generation of tungsten carbide cored ammunition for small arms weapons. This type of ammunition is now available in the 5.56mm, 7.62mm, .338 Magnum and 0.5 inch calibres from companies like Bofors, RUAG and Nammo. The 7.62 round is now a well established munition widely fielded by several Western armies. It is also the armour defining direct fire threat for level 3 of STANAG 4569, the standard that defines the protection levels for light armoured vehicles[1].

The use of tungsten carbide for cores of these modern armour piercing rounds has enhanced the penetration into armour steel by up to 60%. Furthermore the cores are of a similar hardness to many of the usual alumina ceramics used in ceramic armour systems, causing the weight of armour required to defeat such rounds to be significantly increased relative to that necessary to defeat the traditional steel cored AP ammunition. Therefore there is considerable current interest in optimising lightweight ceramic armour solutions to effectively defeat the whole family of tungsten carbide cored bullets.

There are two well established methods of assessing the performance of ceramic armour materials. The first is the gathering of ballistic limit data, where the ceramic is backed with a thin laminar backing, similar to a practical armour design and the velocity of the striking rounds is varied to determine the fifty percentile velocity of complete penetration. (V_{50}). Repeating this test on identical backings for several thicknesses of ceramic will give a ballistic limit curve. However such a procedure uses a significant amount of ceramic material, targets and ammunition and it is therefore a costly procedure to complete.

The second test method is depth of penetration testing (DoP), where the ceramic is backed with a semi-infinite homogeneous material, usually armour steel or aluminium. The target is shot and the residual penetration of the bullet into the semi-infinite backing material is then measured. Whilst the realism of such a test regime can be challenged in terms of formulating a final armour design, DoP testing is an excellent screening test for ceramic material, whereby a large amount of data can be collected quickly and relatively inexpensively from a minimal amount of samples. Whilst relative to any particular practical light armour design perhaps the ranking of ceramic materials by DoP testing may not be exact, however it is generally true that materials failing to show the necessary promise in terms of performance against the particular test projectile in the DoP regime will also perform poorly in the practical armour system.

Ceramics such as alumina, boron carbide, silicon carbide and titanium diboride have long been used in ceramic armour systems. Aluminim nitride and more especially silicon nitride have also found niche applications. Senf et al[2] studied two different grades of alumina ceramic, 92 and 99.5% showing the significantly enhanced performance of the higher grade ceramic against the same type of tungsten carbide cored projectile used for the work described below. However this work also indicates that even a high specification alumina material is unlikely to deliver a very lightweight armour system against such penetrators.

Fully dense boron carbide, silicon carbide, and titanium diboride materials are all much harder (Hv >2000) than alumina (max achievable Hv ~1650). As such the performance of these non-oxide ceramics against tungsten carbide cored projectiles is of significant interest. Silicon nitride generally has a higher toughness value (K_{1C} ~6 MPa m$^{\frac{1}{2}}$) than either silicon carbide or boron carbide (K_{1C} ~4 MPa m$^{\frac{1}{2}}$) and it was felt that this might be significant in determining ceramic performance.

EXPERIMENTAL

The Depth of Penetration (DoP) technique as described by Rozenberg and Yeshurun[3] was used to measure the ballistic performance of the ceramic tiles. For backing for the DoP experiments, a common engineering aluminium alloy Al 6082 T651 was used (YS=250MPa). The test backing were 50.8 × 50.8mm pieces cut from a single 25 mm thick plate. For each ceramic tile of specific thickness (t_c), a single bullet was fired at the target and the residual penetration (P_r) into the aluminium alloy was measured (see Figure 1); at least three experiments were done for each tile thickness.

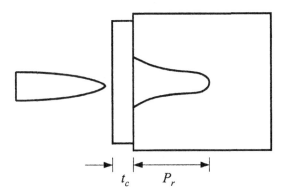

Figure 1: DOP technique for assessing each ceramic's ballistic performance.

Five ceramic materials were compared, four were manufactured by Ceradyne Inc. and one was manufactured by Morgan AM&T. The Ceradyne samples were silicon nitride (Reaction Sintered Ceralloy® 147-31N), Titanium Diboride (Hot Pressed Ceralloy® 225), boron carbide (Hot Pressed Ceralloy® 546-3E) and silicon carbide (Hot Pressed Ceralloy® 146-1S). Morgan AM&T silicon carbide (Sintered Purebide® PS5000) is added for comparison. Some properties of the ceramics are provided below in Table I.

Each ceramic tile was glued to the aluminium alloy backing block using Araldite 2015. This was applied to the mating surfaces and then the ceramic and aluminium block were pushed together and oscillated until an even thin adhesive line had been achieved with no gaps or obvious air inclusions.

Table I.

	Si$_3$N$_4$*	TiB$_2$*	B$_4$C**	Ceradyne SiC**	Morgan SiC*
Density (Kg/m^3)	3100	4500	2500	3150	3140
Hardness (VHN)	1793(2.0)	2226(2.0)	3200	2300	2644(2.0)

* Hardness measured using an Indentec HWDM7 Digital Micro Hardness Machine.
** Hardness value from manufacturer's data.

The range set up was one of a fixed test barrel mounted ten metres from the target. Bullet velocity was measured using the normal sight-screen arrangement. The test ammunition was 7.62 × 51 mm NATO FFV ammunition was used as factory loaded and generated a mean velocity of 973m/s. The bullet core consists of tungsten carbide core (composition by percentage weight C 5.2, W 82.6, Co 10.5, Fe 0.41) of hardness 1200Hv, mounted in a low carbon steel jacket with gilding metal, on an aluminium cup[4] (see Figure 2 below).

Figure 2: 7.62 × 51mm FFV bullet and core.

The test jig was firmly clamped to a test fixture adjustable for height and lateral position and axially aligned with the direction of shot. The jig position was accurately adjusted to ensure that the centre of the target block corresponded with the centre of the shot-line; the jig used engineering vee-blocks as clamping elements. Each of the samples was clamped in place in turn with the ceramic sample protruding out of the front of the clamps. Behind the sample, in the vee-blocks, were three more of the 25 mm blocks of aluminium giving a possible total DoP of 100 mm – effectively semi infinite for the purposes of the test ammunition.

After testing the aluminium alloy blocks were X-rayed allowing the residual penetration to be accurately measured. Furthermore the level of fragmentation of the core and the overall shape of the penetration crater was assessed from the X-rays.

Ceramic Armor and Armor Systems

RESULTS

Figure 3 summarizes the performance of each grade of ceramic. The areal densities of the aluminium alloy penetrated and the ceramic tile thickness are plotted to provide comparison with other published data.

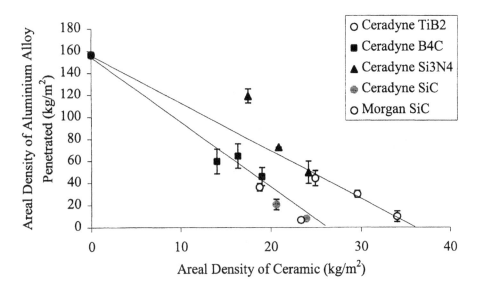

Figure 3: Reduction in areal density of Al6082 T651 penetrated with increasing ceramic areal density.

Linear lines of regression are fitted to the boron carbide and titanium diboride results; the silicon carbide results are taken from [5] (for 6.0–7.5mm tile thickness). Both the hot pressed Ceradyne and sintered Morgan materials have similar performance. The mean result for each thickness with the standard error is presented.

The critical thickness required to stop penetration for each of the four Ceradyne ceramics tested was derived by extrapolating the lines of regression to the point where no aluminium alloy was penetrated (Figure 3). For a similar silicon carbide, titanium diboride and boron carbide the established critical thickness' are similar to the thickness' obtained by Rozenberg and Yeshuran[3] for 12.7 and 14.5mm steel cored bullets. This indicates the relatively high performance of the 7.62 × 51mm FFV bullet. The critical thickness' obtained in this experimental programme are as follows:

Table II.

Ceramic	Critical thickness (mm)
silicon carbide[5]	8.5
titanium diboride	8.0
boron carbide	10.5
silicon nitride	11.3

DISCUSSION

In comparing the different ceramic materials it is clear that silicon carbide and the boron carbide out perform the titanium diboride and silicon nitride on a weight-by-weight basis. There appears to be a two-tier order in the ballistic performance of ceramic against this bullet. It is interesting to note that silicon carbide and boron carbide behave similarly when comparing the areal densities of the ceramic despite the boron carbide possessing a significantly higher hardness and lower density. Figure 4 shows the effect of the varying hardness of the different (Ceradyne) ceramic tiles on the ballistic performance of each tile for a constant tile thickness (6.5mm). For the same thickness, the boron carbide performs worse than the titanium diboride and the silicon carbide tiles.

Furthermore, on comparing the X-rayed witness blocks it was observed that the in some cases, the WC core projectile was not shattered despite perforating the boron carbide of tile thickness in the range 5.6 – 7.6mm (see Figure 5), while all the cores that perforated the silicon carbide tiles were completely fragmented in a similar tile thickness range.

Recently, Moynihan et al[6] presented some results that suggested that the unexpected poor performance of the boron carbide with this type of round is due to the magnitude of shock stresses that occur during penetration of the ceramic by a WC-Co projectile. Using uniaxial strain compressive shock data for tungsten carbide from Grady[7] (5.7% Co, 1.9% Ta) and calibrated Johnson-Holmquist constitutive models for boron carbide (JH-2)[8] and silicon carbide (JH-1)[9] we have run some two-dimensional axially symmetrical hydrocode simulations in AUTODYN™[10] of the WC-Co projectile impacting both ceramics. The calculations show that under uniaxial strain loading conditions the hydrostatic stresses that occur in the ceramics are 18.8GPa for the boron carbide and 21.0GPa for the silicon carbide for a velocity of impact of 973m/s. For the complex loading condition that occurs when the FFV core penetrates the ceramic, the shock stresses were much higher – with values in excess of 40 GPa being recorded in these numerical simulations. However, the simulated deviatoric response (strength vs. hydrostatic stress) of boron carbide was based on an extrapolation beyond relatively lower pressures than this (~ 4GPa) and therefore caution is exercised when interpreting this latter result.

Feng *et al* [11] loaded a Cercom silicon carbide under plain strain loading conditions to peak longitudinal stresses ranging from 10-24GPa. The strength (von Mises equivalent stress) for this silicon carbide prior to failure increased to a maximum of ~13.8GPa; extrapolation of the available data suggested that gradual softening occurred when the hydrostatic stress was increased. With boron carbide under plane strain loading conditions it has been observed that this 'softening' for the pre-failed material occurs in a more dramatic fashion[12]. Furthermore, the reported strength of the un-failed maximum strength of boron carbide is lower than that presented by Feng *et al*. This pressure 'softening' of boron carbide reduces its ability to resist the FFV core when penetrating and therefore ballistically on a thickness basis it performs worse than the silicon carbide.

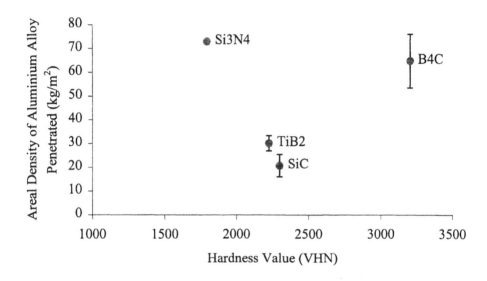

Figure 4: The effect of the hardness value on the ceramic's ballistic performance (measured in the areal density of the aluminium alloy penetrated) for a 6.5mm thick tile. All ceramics are manufactured by Ceradyne Inc.

In comparing the effect of each ceramic on the WC-Co core of the bullet, the amount of fragmentation in the core was inversely proportional to the amount of aluminium alloy that the core penetrated. With the silicon nitride, the data for the 5.5 mm (17.5kg/m^2) tile thickness suggest that its ballistic performance (i.e. its ability to reduce penetration into the aluminium alloy) is less efficient than thicker tiles. This is because the thinner tiles were unable to induce sufficient fracture in the core (see Figure 6a) and is probably caused by the lack of confinement offered by a thin and relatively soft tile. Furthermore, the state of the core after penetration varied from intact (with evidence of tip erosion and fracture across the central trunk of the core) to extensive fragmentation when increasing the tile thickness from 5.5mm to 7.5mm (see Figure 6b).

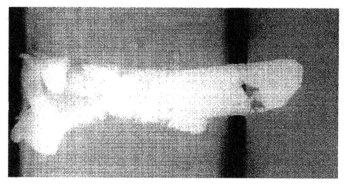

Figure 5: The WC core of the bullet after perforating the 6.5mm of boron carbide and penetrated 2 × 25mm aluminium alloy plates.

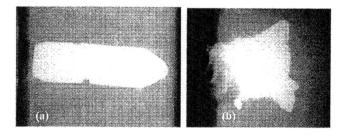

Figure 6: The core in the aluminium alloy after completely penetrating (a) 5.5mm and (b) 7.5mm of silicon nitride.

The mechanism that causes the significantly increased fragmentation of the FFV core by the addition of just 2.0mm of silicon nitride is still not well understood at this stage but it may be to do with the magnitude and duration of the shock stress (increasing the thickness of the ceramic tile, increases the duration of time that the core is loaded in compression). The Ceradyne silicon carbide has been shown to be more sensitive to this effect with the core just fractured but remaining together after completely penetrating a tile thickness of 5.5mm and the core completely fragmented and dispersed with a tile thickness of 6.5mm[5] – again, this would suggest that the magnitude of the shock stress incurred by the core is important in breaking it up. With the titanium diboride targets, the core was fractured for all thickness' (5.5 - 7.5mm); the depth of penetration into the aluminium alloy occurred due to the kinetic energy of the fragments. For thickness' of 6.5mm and above, a very small mass of core fragments were left in the aluminium alloy witness block (see Figure 7).

Ceramic Armor and Armor Systems

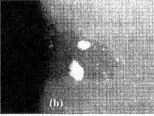

Figure 7: The remnants of the core in the aluminium alloy after completely penetrating (a) 5.5mm and (b) 6.5mm of titanium diboride (penetration into the aluminium alloy was minimal for the 7.5mm thick tile).

There is a marked difference in fracture toughness values between the silicon carbide and silicon nitride ceramics used in this work. The work confirms the importance of adequate ceramic hardness over toughness as being the performance driver against the FFV ammunition for this DoP armour test configuration. Titanium diboride is remarkable in this series of ceramics in both being simultaneously hard and relatively tough (K_{1C}=5.47 MPa m$^{\frac{1}{2}}$) and in terms of performance vs thickness had the overall best result. The relationship between the materials properties of the penetrator core and the ceramic in the armour has been previously reviewed[13] and perhaps in the case of this type of ammunition can be elucidated by a further programme of work.

CONCLUSIONS

As would be expected, increasing the hardness of the ceramic tile offered to the tungsten carbide cored round resulted in an increase of armour performance. That is, except for the boron carbide ceramic which under performed relative to its hardness. This experimental programme showed that the areal density required to provide protection against the round was comparable to that of silicon carbide. However the thickness required to induce comparable fragmentation in the round was higher. This adds to the growing body of evidence that this ceramic does not perform as well as could be expected for this round[14]. The good overall performance of silicon carbide ceramics against this round has confirmed their potential use in practical armour solutions against this ammunition.

With the ceramics tested it was observed that the penetration mechanisms that occur change from that of a rigid body penetration to 'broken body' penetration with the addition of a relatively thin amount of ceramic. This would suggest that not only magnitude of the shock stress is important to the failure of the core of the projectile but also the time of loading. The influence of the fracture toughness of the ceramic on the magnitude and duration of this shock stress may be a suitable subject for further work.

ACKNOWLDEGEMENT
The authors particularly wish to thank Ceradyne Corp of Costa Mesa CA USA and Morgan AM&T of St Marys PA USA, for providing ceramic samples used in this work. Also Steve Champion and Paul Moth of Cranfield University, Royal Military College of Science Shrivenham for their assistance with ballistic testing and hardness measurements.

REFERENCES

[1] STANAG 4569 Edition 1 – Protection Levels for Occupants of Logistic and Light Armoured Vehicles North Atlantic Treaty organisation (NATO) Standardisation Agreement, AC/225-D/1463, 8 March 1999

[2] H. Senf, E. Straβburger, H. Rothenhäusler, and B. Lexow, "The Dependency of Ballistic Mass Efficiency of Light Armor on Striking Velocity of Small Caliber Projectiles" in the *Proceedings of the 17th International Symposium on Ballistics*, Midrand, South Africa, 23-27 March (1998).

[3] Z. Rozenberg and Y. Yeshurun, "The Relationship between Ballistic Efficiency and Compressive Strength of Ceramic Tiles," *Int. J. Impact Engng*, **7** [3] 357-62 (1988).

[4] M.R. Edwards and A. Mathewson "The Ballistic Properties of Tool Steel as a Potential Improvised Armour Plate," *Int. J. Impact Engng*, **19** [4] 297-309 (1997).

[5] C. Roberson and P.J. Hazell, *ibid.*

[6] T.J. Moynihan, J.C. LaSalvia and M.S. Burkins "Analysis of Shatter Gap Phenomenon in a Boron Carbide / Composite Laminate Armor System,". In the *Proceedings of the 20th International Symposium on Ballistics*, Orlando, FL, 23-27 September 2002.

[7] D. Grady, "Impact Failure and Fragmentation Properties of Tungsten Carbide," *Int. J. Impact Engng* **23** 307 – 17 (1999).

[8] G.R. Johnson and T.J. Holmquist, "Response of Boron Carbide subjected to Large Strains, High Strain Rates and High Pressures," *Journal of Applies Physics* **85** [12] 8060-73 (1999).

[9] T.J. Holmquist and G.R. Johnson, "Response of Silicon Carbide to High Velocity Impact" *Journal of Applied Physics* **91** [9] 5858-66 (2002).

[10] Century Dynamics Ltd, Dynamics House, Hurst Road, Horsham, West Sussex, RH12 2DT, UK.

[11] R. Feng, G. F. Raiser and Y. M Gupta, "Material Strength and Inelastic Deformation of Silicon Carbide Under Shock Wave Compression," *Journal of Applied Physics* **83** [1] 79-86 (1998).

[12] N.K. Bourne and G.T. Gray III, "On the Failure of Boron Carbide under Shock"; pp. 775-8 in *Shock Compression of Condensed Matter-2001* edited by M.D. Furnish, N. N. Thadhani and Y. Horie, American Institute of Physics (2002).

Ceramic Armor and Armor Systems

[13] C. Roberson, "Ceramic Materials and their use in Lightweight Armour Systems," Lightweight Armour Systems Symposium, Royal Military College of Science, Shrivenham, UK. June 28-30 (1995).

[14] W. Gooch and M. Burkins, "Dynamic X-Ray Imaging of Tungsten Carbide Projectiles Penetrating Boron Carbide," American Ceramic Society, 27th Annual Cocoa Beach Conference, January 26-31, (2003).

RESISTANCE OF SILICON CARBIDE TO PENETRATION BY A TUNGSTEN CARBIDE CORED PROJECTILE

C. Roberson
Advanced Defence Materials Ltd.
Rugby
Warwickshire.
CV21 3QP, UK.

P. J. Hazell
Cranfield University
Royal Military College of Science
Shrivenham, Oxfordshire
SN6 8LA, UK.

ABSTRACT

Silicon carbide is well known as being a ceramic that can be employed as part of an effective armour solution to defeat small arms ammunition. Its relatively high hardness and ability to accommodate large plastic strains at high confining pressures lend itself to offering sufficient resistance to defeat tungsten carbide cored projectiles. In this paper, the 7.62 × 51mm FFV[*†] round consisting of a tungsten carbide core (Hv 1200) and copper gilding jacket was fired at a variety of thicknesses and types of silicon carbide. The results suggest that it is not only the type of silicon carbide that is important in the design of the armour solution but also a critical thickness that is necessary to sufficiently resist and damage the round. This paper will also draw conclusions on the effectiveness of the manufacturing routes of silicon carbide for armour solutions. This paper will be of interest to armour system designers and manufacturers.

INTRODUCTION

The popularity of silicon carbide for use in lightweight armour systems is increasing rapidly. The major driver for this increase in popularity is the significant improvement in cost / performance ratio of silicon carbide ceramics seen in recent years relative to established materials like alumina. Silicon carbide

---

[*] FFV Ordnance now Bofors Carl Gustav AB.
[†] US DoD designation M993.

---

ceramics are available in a number of compositions from several processing routes, all with varying cost and associated performance levels.

Traditionally high performance silicon carbide armour ceramics have come from a hot pressing route. However such ceramics are costly to make and the capacity for their production is necessarily limited[1]. The direct sintering route is also well established, furthermore there has been a dramatic reduction in the cost of starting powders for such materials over the past ten years leading to a significant improvement in cost competitiveness relative to alumina ceramics. However in the past, sintered silicon carbide in armour applications has generally been characterised as being expensive and brittle with poor tile corner performance and thereby of modest interest in light armour designs relative to alumina. Whilst diversity of composition has developed in other technical ceramics like alumina this was somewhat inhibited in silicon carbide ceramics by several patent law cases. Now, fortunately, some of the patents have lapsed and there is some real optimism in the armour community that new tougher grades of sintered silicon carbide armour ceramics will emerge that will rival the performance of the hot pressed material, but at a competitive cost[2]. Such developments are imperative if the armour on the next generation of highly protected rapidly air-transportable armoured vehicles is going to be available in sufficient quantity and have the required performance at an affordable price[3].

The other route to silicon carbide armour ceramics is the reaction bonding or reaction infiltration route, whereby a compact of relatively coarse silicon carbide powder often containing a carbon binder, is infiltrated with molten metal in a vacuum furnace. The metal used is most often silicon and in the presence of the carbon binder there is a secondary reaction to form a fine silicon carbide matrix phase. As a rule there will be excess un-reacted metal and as such the result has the general characteristics of a mixed ceramic /metal matrix composite. Other metals like aluminium have been used for infiltration and there is the potential for alloying the infiltrating metal as well. Other particulate materials like boron carbide are potentially also possible[4].

Depth of Penetration (DoP) testing is a highly effective method of assessing relative ceramic performance. This type of testing is highly efficient in terms of materials use and time. Whilst the DoP test method can be challenged in respect of the realism relative to practical light armour systems, DoP can be very effectively used to grade the performance of candidate ceramics against particular projectiles. The work reported below is an initial screening test on several commercially available silicon carbide ceramics relative to their performance against the popular tungsten carbide cored 7.62 Bofors FFV round (US code M993).

EXPERIMENTAL

The DoP technique as described by Rozenberg and Yeshurun[5] was used to measure the ballistic performance of the ceramic tiles. For backing for the DoP experiments, a common engineering aluminium alloy Al 6082 T651 was used

Ceramic Armor and Armor Systems

(YS=250MPa). The test backing were 50.8 × 50.8mm pieces cut from a single 25 mm thick plate. For each ceramic tile of specific thickness (t_c), a single bullet was fired at the target and the residual penetration (P_r) into the aluminium alloy was measured (see Figure 1); at least three experiments were done for each tile thickness.

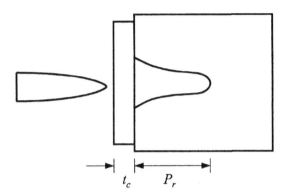

Figure 1: DOP technique for assessing each ceramic's ballistic performance.

The silicon carbide ceramics used in these trials were as follows: Ceradyne Ceralloy® 146-1S which is a material manufactured by the hot pressing route, Morgan AM&T Purebide® PS 5000 and Wacker-Chemie SiC 100 both of which are sintered materials with a boron /carbon sintering aid system, then AME silicon carbide a liquid phase sintered material with an alumina / magnesia precursor system containing 86% silicon carbide content. The final material was Morgan AM&T reaction bonded silicon carbide (Purebide R) which is a classic reaction bonded material made by infiltration of molten silicon into a carbon containing compact of silicon carbide grain under in a vacuum furnace at around 1600 degree Celsius. The resultant material contains approximately 10% un-reacted silicon metal. The ceramic tiles had approximately the plan view dimensions of 50.0 × 50.0 mm and were within a thickness range 5.5 to 8.0 mm. Originally the ceramics were supplied either as 101.6 × 101.6mm or 50.0 × 50.0mm tiles and where necessary were diamond cut to provide suitable test samples for this work. Some properties of the ceramics are provided below in Table I.

Table I. Density and Hardness of the Ceramic Materials Used.

	Ceradyne*	Morgan (Sint.)**	AME**	Wacker**	Morgan (RB)**
Density (kg/m³)	3150	3140	3220	3140	2980
Hardness (VHN)	2300	2644(2.0)	2228(2.0)	2637(2.0)	1975(0.5)

* Hardness value from manufacturer's data.
** Hardness measured using an Indentec HWDM7 Digital Micro Hardness Machine.

The adhesive used was Araldite 2015. This was applied to the mating surfaces and then the ceramic and aluminium block were pushed together and oscillated until an even thin adhesive line had been achieved with no gaps or obvious air inclusions.

The range set up was one of a fixed test barrel mounted ten metres from the target. Bullet velocity was measured using the normal sight-screen arrangement. The test ammunition was 7.62 × 51 mm NATO FFV ammunition was used as factory loaded and generated a mean velocity of 973m/s. The bullet core of the FFV ammunition consists of tungsten carbide core (composition by percentage weight C 5.2, W 82.6, Co 10.5, Fe 0.41) of hardness 1200Hv, mounted in a low carbon steel jacket with gilding metal, on an aluminium cup[6] (see Figure 2 below).

Figure 2: 7.62 × 51mm FFV bullet and core.

The test jig was firmly clamped to a test fixture adjustable for height and lateral position and axially aligned with the direction of shot. The jig position was accurately adjusted to ensure that the centre of the target block corresponded with the centre of the shot-line; the jig used engineering vee-blocks as clamping elements. Each of the samples was clamped in place in turn with the ceramic sample protruding out of the front of the clamps. Behind the sample, in the vee-blocks, were three more of the 25 mm blocks of aluminium giving a possible total DoP of 100 mm – effectively semi infinite for the purposes of the test ammunition.

After testing the aluminium alloy blocks were x-rayed which allowed the residual penetration to be accurately measured. Furthermore the level of fragmentation of the core and the overall shape of the penetration crater was assessed from the x-rays.

Ceramic Armor and Armor Systems

RESULTS

Figure 3 below summarizes the performance of each grade of silicon carbide. The areal densities of the aluminium alloy penetrated and the ceramic tile thickness are plotted to provide comparison with other published data. The standard error of the mean of the penetrated areal density is presented.

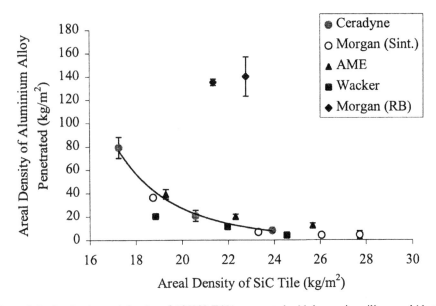

Figure 3: Reduction in areal density of Al6082 T651 penetrated with increasing silicon carbide tile areal density.

The reaction bonded material was markedly different from the rest of the ceramics and performed rather poorly. In these trials, it did not break the core of the projectile. All of the other four grades performed similarly. As the areal density of the ceramic is increased, the depth of penetration into the Al6082 T651 is reduced. With the four better performing ceramics, there were four types of characteristic craters that were observed and represent the different stages of round break up that occur: For areal densities of ceramic less than 18kg/m² (Ceradyne), a narrow crater similar to the diameter of the core of the bullet (5.59mm) was formed; the core was fragmented but not dispersed (see Figure 4). For areal densities in excess of 18kg/m² no evidence of the core was retrieved. This was due to extensive fragmentation and dispersion of the core. For areal densities in the range of 18kg/m² to 20kg/m² a fat neck with a relatively thin crater was evident. For 20k/m² to 26kg/m², a shallow yet relatively wide crater was formed and finally, with an areal density in excess of 26kg/m² the round was overmatched leaving a small indentation in the aluminium alloy witness block.

We can re-plot the above data to assess the added advantage of increasing the thickness of the ceramic on the ballistic performance (taking into account the added mass) of the complete armour system. The ballistic performance of the armour is therefore calculated from:

$$MEF = \frac{\rho_b P_b}{\rho_c t_c + \rho_b P_r} \tag{1}$$

where MEF is a mass efficiency factor, t is the thickness, ρ is the bulk density, and P is the penetration. Subscripts b and c represent the block and the ceramic respectively; P_r is the residual penetration into the witness block when there is a ceramic plate present (see Figure 5).

The MEF for the Morgan AM&T Purebide R is not plotted in Figure 5. It was found to have a value of 1.0 indicating that ballistically, it was no more effective than the aluminium alloy. Furthermore, a second order polynomial trend line is fitted through the Ceradyne data. The asymptotic trend of the data suggests that as the thickness of the ceramic is increased, the added advantage of increasing the thickness further is reduced until a critical thickness is reached.

The above factor is quite useful in determining the efficiency of a particular ceramic tile. For example, the Ceradyne silicon carbide sample has a MEF of 4.9 with a thickness of 7.59mm. This means that, ignoring edge effects and the resistive effect due to longitudinal confinement, we will need 4.9 times the areal density of the aluminium alloy to provide the same protection as the 7.59mm silicon carbide tile bonded to the aluminium alloy.

Figure 4: FFV core before and after perforating 5.44mm of Ceradyne silicon carbide (65% recovered mass).

The critical thickness required to stop penetration for each of the four best performing ceramics tested was derived from Figure 3. They are as follows:

Ceramic Armor and Armor Systems

Table II. Critical Thickness Of Ceramic to Prevent Penetration into the Aluminium

Ceramic	Critical thickness (mm)
Wacker	8.4
Ceradyne	8.5
Morgan (Sint.)	8.5
AME	9.0

No critical thickness was established for the Morgan AM&T Purebide R.

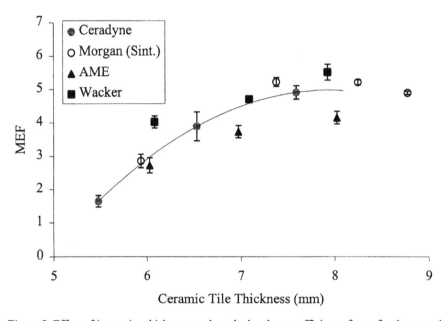

Figure 5: Effect of increasing thickness on the calculated mass efficiency factor for the ceramic.

DISCUSSION

In spite of a similar hardness value to the AME material the reaction bonded material had little interaction with the projectile and this is indicative of a completely different ceramic fracture mechanism operating with this material. The poor performance is attributed to the significant un-reacted silicon content. With the exception of the reaction bonded material the different silicon carbides behaved similarly in defeating the core however, the above results indicate that the harder ceramics (Wacker and Morgan) performed slightly better with this round; Wacker SiC 100 performed slightly better than the other ceramics throughout the thickness range despite an inability to sustain shear strength when failed under high confining pressures[7]. Aside from reaction bonded the softer AME SiC consistently performed the worst.

The thicker tiles of ceramic were more efficient at stopping the round than thinner tiles. What is remarkable is that even with a relatively small addition to the tile thickness (~1.5-2.0mm) there is a transition from effectively a rigid body penetration where the core has fractured but remains together, to 'broken body' penetration where the core fractures and the fragments are dispersed during penetration. This is illustrated below with the post firing X-rays of the aluminium alloy blocks with two of the Ceradyne results (Figure 6). With a 5.48mm tile, the core completely penetrates the tile and continues to penetrate into the aluminium alloy despite suffering fracture resulting in a crater that is long and thin. Increasing the thickness of the tile by 1.04mm, results in the core being fragmented and dispersed. The core fragments penetrate into the aluminium alloy resulting in a wide and shallow crater. This would suggest that for this particular tungsten carbide cored projectile that fracture and dispersion of the core is dependent on the time of its contact with ceramic and therefore compressive loading during penetration. This would suggest that a crack softening approach where the gradual failure of the material is simulated is appropriate for modelling the failure of this tungsten carbide core.

Unfortunately, no firings were done with this experimental set-up where a critical thickness that causes macroscopic fracture in the core was established. However, firings into relatively softer (1975Hv [0.5]) but thicker (>7.00mm) reaction bonded silicon carbides resulted in little or no fracture in the core (see Figure 7). Furthermore, previous work[8] with a relatively soft 92% grade alumina has shown that a tungsten carbide cored projectile completely penetrates in a rigid manner for tile thickness' up to 30mm and for velocities less than 1000m/s. Increasing the velocity of impact to 1200m/s, a transition region occurred where the penetration changed from being rigid body to 'broken body' where the penetrator was comminuted. This would suggest that the magnitude of the shock stress from either a higher compressive strength or increased velocity leads to the onset of the tip crushing and fracture of the core body.

Figure 6: X-ray of the depth of penetration into the aluminium alloy witness block after completely penetrating (a) 5.48mm and (b) 6.52mm of Ceradyne silicon carbide (Ceralloy® 146-3E).

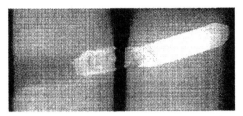

Figure 7: X-ray of the depth of penetration into the aluminium alloy witness block after completely penetrating 7.16mm of Morgan Reaction Bonded silicon carbide. Note that the core remains in tact.

CONCLUSIONS

Comparing the silicon carbides showed that the harder ceramics performed better. Furthermore, the ceramic system can be optimised by the addition of a relatively small increase in thickness of silicon carbide. The addition of the relatively small thickness of material results in a transition from rigid body penetration to complete fragmentation of the core and as a result broken body penetration. If during the penetration of the ceramic there is sufficient time so that the cracks in the core are able to grow so that the core structure is completely compromised the fragments are dispersed. If, however, the complete fracture of the core occurs *after* the complete penetration of the ceramic, the relatively soft aluminium alloy does not disperse the core fragments.

ACKNOWLDEGEMENT

The authors wish to thank Morgan AM&T of St Marys PA USA, Morgan Matroc Limited, Rugby England and Ceradyne Corp of Costa Mesa CA USA for providing ceramic samples used in these tests. We also thank: Steve Champion and Paul Moth of Cranfield University, Royal Military College of Science Shrivenham for their assistance with ballistic testing and hardness measurement.

REFERENCES

[1] R.E. Tressler, "An Assessment of Low Cost Manufacturing Technology for Advanced Structural Ceramics and its Impact on Ceramic Armor," *Ceramic Transactions* **134** pp. 451-62 (2002).

[2] D.A. Ray, R.M. Flinders, A. Anderson and R.A. Cutler, "Hardness / Toughness Relationship for SiC Armor," American Ceramic Society, 27th Annual Cocoa Beach Conference, January 26-31, 2003.

[3] SBIR solicitation: A01-045 Silicon Carbide Based Multiphase Composites for Armor Applications, May 2001.

[4] M.K. Aghajanian, B.N. Morgan, J.R. Singh, J. Mears, R,A, Wolffe, "A New Family of Reaction Bonded Ceramics for Armor Applications," *Ceramic Transactions* **134** pp. 527-39 (2002).

[5] Z. Rozenberg and Y. Yeshurun, "The Relationship between Ballistic Efficiency and Compressive Strength of Ceramic Tiles," *Int. J. Impact Engng*, **7** [3] 357-62 (1988).

[6] M.R. Edwards and A. Mathewson "The Ballistic Properties of Tool Steel as a Potential Improvised Armour Plate," *Int. J. Impact Engng*, **19** [4] 297-309 (1997).

[7] I. M. Pickup and A. K. Barker, "Deviatoric Strength of Silicon Carbide subject to Shock"; pp. 573-6 in *Shock Compression of Condensed Matter-1999* edited by M.D. Furnish, L. C. Chhabildas and R. S. Hixson, American Institute of Physics (2000).

[8] H. Senf, E. Straβburger, H. Rothenhäusler, and B. Lexow, "The Dependency of Ballistic Mass Efficiency of Light Armor on Striking Velocity of Small Caliber Projectiles" in the *Proceedings of the 17th International Symposium on Ballistics*, Midrand, South Africa, 23-27 March (1998).

AN INVESTIGATION INTO FRAGMENTING THE 14.5 MM BS41 ARMOUR PIERCING ROUND BY VARYING A CONFINED CERAMIC TARGET SET-UP

N. J. Woolmore * and P. J. Hazell
Cranfield University
Royal Military College of Science
Oxfordshire
United Kingdom
SN6 8LA

T.P. Stuart
Physical Protection Group
Physical Sciences Department
DSTL, Porton Down, Wiltshire
United Kingdom
SP4 0JQ

ABSTRACT

In this paper, experimental results are presented from an investigation of the parameters of a ceramic-faced armour system that are required to induce damage in a tungsten carbide penetrator. Experiments consisted of firing the 14.5 mm BS41 tungsten carbide cored projectile into various thicknesses and types of alumina and silicon carbide, backed by aluminium alloy semi-infinite witness plate. In each case, the ballistic resistance of the ceramic-faced armour configuration was assessed by the calculation of a differential efficiency factor. The failure of the projectile, depth of penetration and fracture morphology of the ceramic tile are also presented. Conclusions are drawn on the nature of round defeat.

INTRODUCTION

Over the last three decades a wide range of ceramics have been recognised for their outstanding ballistic performance qualities, their high hardness and strength under compressive loading making them prime candidates for use in modern armour systems. Ceramic armours are good at eroding or fragmenting armour piercing (AP) projectiles and spatially spreading the impact energy.

The BS41 projectile has a tungsten carbide (WC) core, which, because of its high hardness and density outperforms the steel cored 14.5 mm B32 round when penetrating ceramic-faced targets. No single mechanism dominates the dynamic failure of the armour system or the projectile during penetration. This investigation sets out to try to determine what factors influence damage phenomena in WC AP penetrators.

EXPERIMENTAL PROCEDURES

The purpose of this present study is to identify the influence of different armour configurations on their ability to defeat effectively and fragment the WC core of the BS41.

Ammunition

The 14.5 mm BS41 AP Incendiary projectile (Figure 1) was used in the investigation. It has a WC core consisting of W-88%-C-5.58%, Co-5.45% matrix Fe-0.34%, hardness 1289 Hv (2 kgf load).

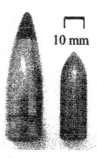

Figure 1. 14.5 mm Armour Piercing/Incendiary Type BS41 projectile, displaying jacketed projectile and core. Further information may be found in Ref. [1].

The BS41 was launched from a fixed test barrel of a KPV anti-tank Russian machine-gun mounted 10 metres from the target. Sky screens and break foils were used to measure the projectile velocity. The mean impact velocity, generated with factory-loaded propellant, was approximately 1016 m/s (± 10 m/s). The propellant mass was varied to produce velocities of approximately 750, 850 and 1100 m/s. Experiments were performed under ambient conditions.

Target Materials

The ceramic targets used in the trial were as follows: Sintox-CL aluminium oxide (98.6 % die pressed and sintered) supplied by Morgan Matroc, silicon carbide AM & T PS 5000 (die pressed and sintered with a boron / carbon sintering aid system) supplied by Morgan AM & T, and silicon carbide B (manufactured by pressure assisted densification) supplied by Cercom USA. All the ceramic tiles were 100 mm x 100 mm x n mm (n ranging from 11 mm to 30 mm). Material properties of the three ceramics used are given in Table I.

Ceramic Armor and Armor Systems

Table I. Material properties of the three ceramics

Material Property	SiC B Cercom Inc.	PS 5000 SiC Morgan AM & T	Sintox-CL Alumina Morgan Matroc
Density (kg/m³)	3180	3140	3890
Hardness (VHN)[1]	1969 (2.0)	2644 (2.0)	1737 (2.0)

[1] Vickers Hardness Results measured using an Indentec HWDM7 Digital Micro Hardness Machine at 2.0 kgf; Cercom's quoted value for the SiC B is 2400 [2], which is somewhat higher than measured in this programme.

The witness plate consisted of an aluminium alloy of yield strength 440 MPa. The aluminium alloy plates were aligned with the short transverse direction of the roll in line with the axis of the penetration. Tungsten 25-micron powder mixed with grease, 0.5 mm thick, was applied between the mating surfaces of the ceramics and witness plates. This was used to achieve a good bonding surface between the tiles with no gaps or obvious air inclusions.

Confinement

The ceramic armour systems were confined in an experimental test jig based on a design presented in Sherman *et al.* [3]. It was positioned on a test fixture axially aligned with the direction of shot (Figure 2). All tests were carried out at normal impact angle, *i.e.*, zero obliquity. A steel containment box with a central hole to allow the projectile to pass through was positioned over the face of the jig to collect all the ceramic and projectile debris. The ceramic systems were firmly clamped in by wedges with bolts tightened to 65 Nm torque.

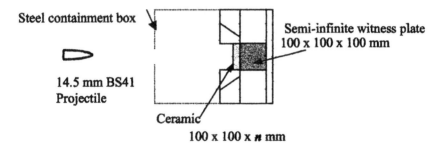

Figure 2. Experimental set up.

Three experiments are presented in this programme of research:

1   Experimental investigation reporting the effects of altering the areal density (ranging from 37 to 98 kg/m²) of three different types of ceramic front plates, Sintox-CL alumina, PS 5000 SiC and SiC B using an aluminium alloy witness plate.

2   Experimental investigation comparing the influence of altering impact velocity between 750 to 1100 m/s. Two different armour systems are reported: 18 mm SiC B and 18 mm Sintox-CL front plate with an aluminium alloy witness plate.

3   Experimental investigation introducing a 10 mm air gap between the aluminium alloy witness plate and a front armour appliqué system. The appliqué system consisted of a ceramic coupled to an aluminium alloy plate using the tungsten paste. Two types of ceramic front plates were compared: 18 mm Sintox-CL and 18 mm SiC B. The thickness of the aluminium alloy plate coupled to the ceramic was varied for each firing (10, 20 and 30 mm).

A minimum of two firings for each experimental set-up was performed to compare data.

Methodology

The DoP technique was employed as an experimental validation to compare the ballistic performance of the model scale targets with the baseline targets. In each case, the ballistic resistance of the ceramic-faced armour configuration was assessed by the calculation of a differential efficiency factor, e_c, [4]. The e_c compares the ballistic performance of the armour system with that of the baseline target and is calculated by equation 1. For a ceramic-faced aluminium alloy target spaced from the semi-infinite witness plate, the equation was modified to take into account the additional areal density offered by the thickness of aluminium alloy ($\rho_{al}.a$) coupled to the ceramic. For a semi-infinite DoP test (Figure 3), $\rho_{al}.a=0$.

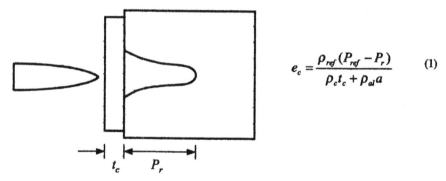

$$e_c = \frac{\rho_{ref}(P_{ref} - P_r)}{\rho_c t_c + \rho_{al} a} \qquad (1)$$

Figure 3. DoP technique with differential efficiency factor equation for assessing each ceramic ballistic performance.

where, P_{ref} and ρ_{ref} are respectively the reference DoP and density of a semi-infinite plate without a ceramic tile, P_r is the residual DoP into reference material after penetration through the armour configuration under test, ρ_c is the density and

t_c is the thickness of ceramic material under test; ρ_{al} is the density of the aluminium alloy plate that was coupled to the ceramic in some experiments.

The projectile core was magnetically separated and weighed. Pulverisation of the mass was assumed with core loss. Ceramic fragments were separated according to size by sifting them through a set of progressively finer sieves. This permitted the WC and ceramic fragment mass distribution and fracture morphology to be analysed.

RESULTS AND DISCUSSION
Performance of Three Different Ceramics

Initial reference firings were used to determine the average baseline DoP value in to different witness plates without a ceramic front plate. This provided data for the e_c to be calculated to evaluate the effectiveness of the different ceramic armour targets.

Figure 4 below summarises the performance of three different ceramic front plates. The areal densities of the aluminium alloy witness plate penetrated and the corresponding ceramic front plate are plotted. To evaluate the critical thickness required to stop the bullet, linear lines of regression were fitted to the data from the three ceramics. Extrapolating the lines of regression to the point where no witness plate was penetrated derived the critical thickness (Table II).

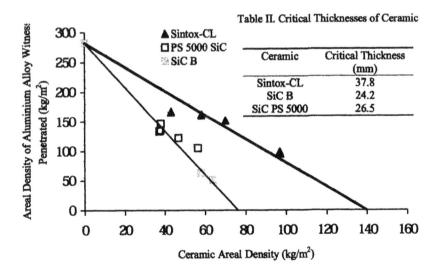

Table II. Critical Thicknesses of Ceramic

Ceramic	Critical Thickness (mm)
Sintox-CL	37.8
SiC B	24.2
SiC PS 5000	26.5

Figure 4. Ceramic areal density of three different ceramics plotted against DoP, velocity of impact is kept constant 1016 m/s (± 10 m/s). Trend-lines are displayed through Sintox-CL and SiC B only.

In comparing the areal density of material penetrated after complete penetration of the SiC B, it consistently out-performed the Sintox-CL. At higher areal densities, the SiC B performed better than the PS 5000. At lower areal densities, the trend line fitted through the SiC B data suggests that the ceramics performed similarly. Extrapolating the lines of regression reveals the SiC B slope diminishes faster than the trend line fitted though the Sintox-CL data. This suggests that it is better at fragmenting and defeating the AP projectile. This was evident from analysis of the WC core retrieved (Figure 6).

(a)

(b)

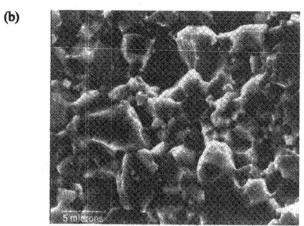

Figure 5. Fractographic images of the comminutia resulting from the ballistic impact tests (a) SiC PS 5000 failed in primarily transgranular cleavage mode (b) SiC B failed in primarily intergranular mode.

Pickup and Barker [5] studied two similar ceramics that had been subjected to 1D stress experiments at strain rates ranging from 10^{-3}/s to 10^{3}/s. One ceramic was SiC B and the other was SiC 100, a pressureless sintered ceramic. In these experiments they measured the time to failure. They noted that the time to failure of the SiC B was 50 % higher than that of the SiC 100 and suggested that this could be explained by the microstructure. The main difference was that there was transgranular cleavage in the SiC 100 and intergranular failure in the SiC B. After comminution, the SiC B consisted of particles of closely interlocked grains. This provided considerable resistance to deviatoric stresses. If the grains cleave (as was observed by the PS 5000 SiC) and co-operative movement is enabled, the shear strength is reduced. Transgranular cleavage failure in the PS 5000 SiC is evident in figure 5. This could suggest, comparing the areal densities of aluminium alloy penetrated with the two different ceramic materials, why the SiC B out-performs the PS 5000 SiC at higher areal densities despite the PS 5000's higher hardness value. With lower areal densities of ceramic, this effect will be less pronounced and therefore the performance was similar (see Figure 4).

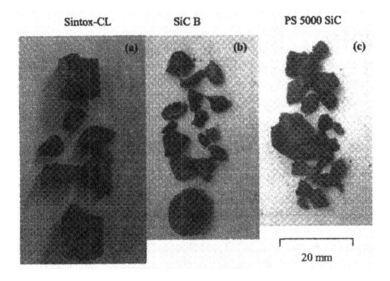

Figure 6. WC fragmentation pictures: (a) 58 kg/m^2 Sintox-CL (88 %) (b) 58 kg/m^2 silicon carbide-B (29 %) (c) 58 kg/m^2 silicon carbide PS 5000 (34 %). (% Total mass recovered)

Figure 6 above compares the WC core fragmentation after penetrating approximately 58 kg/m^2 of Sintox-CL, PS 5000 SiC or SiC B, and an aluminium alloy witness plate. Both Sintox-CL alumina and the two SiC targets interactions caused fracture across the diameter of the core suggesting spall or bending stresses. A greater number of fragments of the WC core were dispersed after impacting a SiC front plate compared with the Sintox-CL. The higher level of fragmentation with SiC is because of its greater strength and hardness that induces a greater shock in to the WC core penetrator. The data suggests that the SiC is good at fragmenting the WC, spatially spreading the impact energy and decreasing the local pressure on the aluminium alloy witness plate.

Performance of ceramic armour system with altered impact velocity

Figure 7 below summarises the performance of Sintox-CL and SiC B when altering impact velocity. The e_c of the ceramic system comprising of two different types of ceramic front plate and an aluminium alloy back plate against impact velocity are plotted to provide comparison.

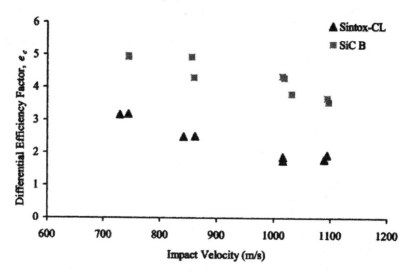

Figure 7. Reduction in e_c of the ceramic armour system. Two different ceramic armour systems (SiC B and Sintox-CL) are plotted against impact velocity.

It is interesting to note that SiC B and Sintox-CL both behaved similarly with a linear decrease in e_c of the ceramic armour system as the impact velocity increases. At both 750 m/s and 1100 m/s there was a 1.6 times increase in e_c of the SiC B ceramic system compared with the Sintox-CL. The rate of decrease of e_c tends to drop for Sintox-CL at higher impact velocities compared to the SiC B. As the DoP increases, the proportional thickness of ceramic drops reducing the e_c of the system.

Evaluating armour systems against projectiles with hard cores at different impact velocities, it is possible to encounter a change in projectile and target defeat mechanisms. The reduction of e_c as impact velocity increased is due to the higher shock stress induced in the ceramic leading to greater fragmentation of the

Ceramic Armor and Armor Systems

ceramic. After each firing, fragments from the target were recovered and sieved into different sizes. It was observed that greater proportions of smaller fragments were recovered at the higher impact velocity.

Examination of recovered WC core and fragmentation behaviour of ceramic revealed a change in fracture morphology with increased impact velocity. Figure 8 below displays the WC core retrieved after firing into two different ceramic front plates at different velocities.

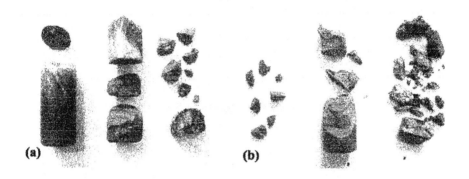

Figure 8. WC core fragmentation for two different types of ceramics front plate, aluminium alloy witness plate (a) Sintox-CL (b) SiC B. From left to right 750 m/s, 850 m/s, 1100 m/s.

Evaluating the WC recovered from firing into Sintox-CL saw a distinct increase in number, and greater pulverisation of fragments with increased impact velocity. SiC B displayed unusual behaviour; at 750 m/s, there was relatively large amount of core comminution when compared to the impact at 850 m/s. The core was shattered in all the velocity regimes.

Introducing an Air Gap in-between the Ceramic Appliqué and the Witness Plate

Having completed our previous experiments we proceeded to compute the e_c for two different ceramic front plates introducing an appliqué armour system (Figure 9).

Different thicknesses of aluminium alloy were coupled to the ceramic. A 10 mm air gap was positioned between the witness plate and the armour appliqué system. Our goal was to evaluate the performance of a spaced aluminium appliqué armour system.

Having generated the polynomial curves for the two ceramic tiles a clear difference in e_c is observed. As we would expect, the differential efficiency of the SiC B was consistently greater than the Sintox-CL.

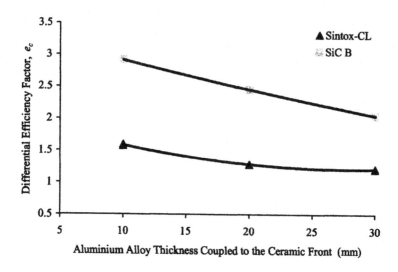

Figure 9. The effect of different aluminium alloy thicknesses coupled to two different ceramic (SiC B and Sintox-CL) front plates with a 10 mm air gap on e_c, velocity of impact is kept constant ± 10 m/s

In comparing the e_c after complete penetration of the SiC B with aluminium alloy coupled plates, it consistently out-performed the Sintox-CL. Sintox-CL displayed a gradual decrease from, 1.6 to 1.2 e_c as the areal density of the aluminium alloy coupled to the ceramic front plate increased. This is unlike SiC B, where the rate of e_c drop with aluminium alloy backing thickness is greater, from 2.9 to 2.0 e_c. This drop is just over two times higher than the change in e_c of the Sintox-CL. The rate of decrease stays fairly linear for SiC backed with an increasing thickness of aluminium, but the rate of decrease decreases for the similar system with Sintox-CL.

The reason for the drop in the value of e_c is due to the ability of the thinner plates of aluminium alloy to deform and bend plastically thereby transferring a greater percentage of the projectile's kinetic energy to a lower form of energy (heat). With thicker aluminium alloy plates, very little bending occurs and the increased mass of the aluminium alloy reduces the value of e_c.

Interestingly at 30 mm of aluminium alloy coupled to the SiC there was no penetration of the witness plate. The WC core was completely dispersed on leaving the front appliqué system not providing enough energy to penetrate the witness plate.

Comparing Figures 7 and 9 shows that the SiC behaves 40 % better when supported with the semi-infinite witness plate when compared to the appliqué system with 10 mm aluminium alloy, the Sintox-CL shows a 25 % improvement. The data indicates that the e_c of both ceramics increases with a high level of rear

Ceramic Armor and Armor Systems

axial support to the tile. This is evident as the targets with 10 to 30 mm coupled aluminium alloy leaves the core largely intact (see Figure 10) while the semi-infinite aluminium backing causes pulverisation of the round.

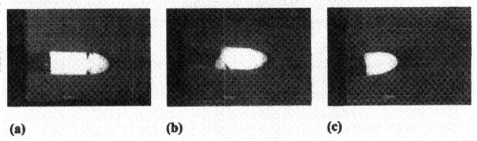

(a)   (b)   (c)

Figure 10. X-rays of the residual penetration and the remnants of the WC core in the aluminium alloy. (a) 10 mm (b) 20 mm (c) 30 mm aluminium alloy coupled to 18 mm Sintox-CL.

Figure 10 above compares the WC tips captured in the aluminium alloy after firing in to the three different ceramic appliqué systems in experiment 3. The residual DoP X-rays show a relatively intact core at lower thicknesses of aluminium alloy backing, where as with the greater thickness backing the aluminium alloy dispersed the rear portion of the fractured core. Greater core loss was due to the formation of the spall crack at different points down the length of the WC core.

CONCLUSIONS

The aim of the experimental programme was to determine the nature of WC failure when penetrating different ceramic armour systems. In comparing the differential efficiency factor of the three different ceramic materials, the SiC B targets out-performed the PS 5000 SiC and the Sintox-CL alumina.

The ceramic front plate plays a significant part in the failure of the WC. Both SiC and Sintox-CL target interactions caused fracture across the diameter of the core suggesting spall or bending stresses. The SiC with its greater strength and hardness induced a greater shock imparting greater fragmentation in the WC core round. The results suggested that at the higher areal densities, SiC B outperformed PS 5000 SiC. The reason for the increased performance of the SiC B over PS 5000 SiC, despite having a lower hardness, is due to the SiC B failing by intergranular fracture and thereby forming particles of closely interlocked grains. These particles provide considerable resistance to flow and therefore the failed ceramic maintains some of its strength under load.

Evaluating armour systems against projectiles with hard cores at different impact velocities, it is possible to encounter a change in projectile and target defeat mechanisms. The reduction of e_c as impact velocity increased is due to the higher shock stress induced in the ceramic. Post-test target analysis revealed that the change in the ballistic behaviour and the WC fragmentation was affected by velocity change.

Data indicates that e_c of the ceramic appliqué decreases with the thickness of rear axial support to tile until a sufficient thickness of material retains the ceramic fragments in front of the projectile. This is evident as 10 to 30 mm aluminium alloy coupled to the ceramic front plate leaves the core largely intact while the semi-infinite aluminium backing causes pulverisation of the round.

ACKNOWLEDGMENTS

The authors are most grateful for the support of the sponsors of DSTL for many helpful discussions concerning the trials, most notably Bryn James, Andy Baxter and Ross Jones. Moreover, the authors wish to thank Colin Roberson for his support in both supplying materials and discussion. We particularly would like to thank Morgan AM & T for supplying the PS 5000 samples, Morgan Matroc Ltd. for supplying the Sintox-CL and Cercom for supplying the SiC B. Finally we would like to thank the tungsten powder suppliers, HC Stark all for making the experimental investigation possible.

REFERENCES

[1] Jane's Ammunition Handbook, 14.5 × 114 mm Armour Piercing Projectile, 1998 - 1999.

[2] Cercom's website, http://www.thomasregister.com/olc/cercom/scb.htm

[3] D. Sherman and T. Ben-Shushan, "The Ballistic Failure Mechanisms and Sequence in Confined Ceramic Tiles," Israel Institute of Technology, TNS Meeting, Pitzbourg, 1993.

[4] R. R. Franzen, D. L. Orphal and C. E. Anderson, "The Influence of Experimental Design on Depth-of-Penetration (DOP) Test Results and Derived Ballistic Efficiencies," International Journal of Impact Engineering, Vol. 19, No. 8, pp. 727 - 737, 1997.

[5] I. M. Pickup and A. K. Barker, "Deviatoric Strength of Silicon Carbide Subject to Shock," Shock Compression of Condensed Matter, pp. 573 - 576, 1999.

Ceramic Armor and Armor Systems

KEYWORD AND AUTHOR INDEX